SHORTWAVE LISTENING HANDBOOK

Harry L. Helms

PRENTICE-HALL, INC., Englewood Cliffs, New Jersey 07632

Library of Congress Cataloging-in-Publication Data

Helms, Harry L. (date)
  Shortwave listening handbook.

  Bibliography. p.
  Includes index.
  1. Radio, Short wave—Receivers and reception.
  2. Radio stations, Short wave—Directories.
  I. Title.
  TK6564.S5H45 1987   384.54′53   86-22617
  ISBN 0-13-809617-1
  ISBN 0-13-809591-4 (pbk.)

Editorial/production supervision and
  interior design: *Nancy Menges*
Cover design: *Wanda Lubelska*
Manufacturing buyer: *S. Gordon Osbourne*

© 1987 by Prentice-Hall, Inc.
A Division of Simon & Schuster
Englewood Cliffs, New Jersey 07632

All rights reserved. No part of this book may be
reproduced, in any form or by any means,
without permission in writing from the publisher.

Printed in the United States of America

10 9 8 7 6 5 4 3 2 1

ISBN 0-13-809617-1 025 {C}
ISBN 0-13-809591-4 025 {P}

PRENTICE-HALL INTERNATIONAL (UK) Limited, *London*
PRENTICE-HALL OF AUSTRALIA PTY. Limited, *Sydney*
PRENTICE-HALL CANADA INC., *Toronto*
PRENTICE-HALL HISPANOAMERICANA, S.A., *Mexico*
PRENTICE-HALL OF INDIA PRIVATE LIMITED, *New Delhi*
PRENTICE-HALL OF JAPAN, INC., *Tokyo*
PRENTICE-HALL OF SOUTHEAST ASIA PTE. LTD., *Singapore*
EDITORA PRENTICE-HALL DO BRASIL, LTDA., *Rio de Janeiro*

This book is dedicated to Gerry Dexter, Don Jensen, Tom Kneitel, and C.M. Stanbury II, who have done much over the years through their writings to promote shortwave listening.

CONTENTS

PREFACE ix

1 WHAT IS SHORTWAVE LISTENING? 1
  Shortwave Broadcasters 3
  Amateur "Ham" Radio 10
  Utility Stations 11
  AM Broadcast Band DXing 13
  Clandestine, Pirate, and Illegal Radio 15
  The Shortwave Listening Hobby 17
  Terminology and Timekeeping 18

2 UNDERSTANDING THE SHORTWAVE SPECTRUM 21
  Modes of Emission 22
  Primary Shortwave Frequency Bands 27

3 SELECTING A SHORTWAVE RECEIVER 40
  Receiver Basics 40
  Frequency Coverage 42
  Frequency Readout and Tuning 43
  Sensitivity and Selectivity 45
  Noise Limiters and Blankers 52
  Automatic Volume and Gain Controls 53
  S-Meters and Signal Indicators 53
  Memories and Multiple Tuning Circuits 54

Beat-Frequency Oscillators and SSB Reception 55
Exalted Carrier SSB Reception 55
Learning to Use a Shortwave Receiver 57
Which Receiver to Choose? 58

4 ANTENNAS AND ACCESSORIES 60

Some Simple Antenna Theory 60
A Simple 1.6–30-MHz Outdoor Antenna 63
Coaxial Cable, Grounds, and Lightning Arrestors 66
The Dipole Antenna 68
Other Types of Antennas 73
Antennas for Limited Space and Difficult Situations 74
Active Antennas 76
Preamplifiers and Preselectors 77
Antennas for the Broadcast Band and Longwave 78
Audio Filters 79
Headphones 80
Radioteletype Receiving Equipment 81

5 RADIO PROPAGATION 82

The Nature of the Ionosphere 84
The Sun and Radio Reception 85
Propagation Paths 88
Refraction Angles and Skip Zones 92
Reception Patterns 94
Propagation Forecasting 96

6 MAJOR INTERNATIONAL SHORTWAVE BROADCASTERS 99

The British Broadcasting Corporation (BBC) 100
Radio Moscow 103
Radio Nederland 108
Radio France Internationale 110
Radio Beijing 112
Radio Tirana 114
Evangelical Broadcasting: HCJB and Trans World Radio 114
Vatican Radio 117
Radio Canada International 119
WRNO, KYOI, and American Commercial Shortwave Broadcasting 120
Radio RSA 122
Deutsche Welle 123
Voice of America 124
Other International Broadcasters 126

Contents   vii

7 DOMESTIC SHORTWAVE BROADCASTING   **127**

Europe   130
USSR Domestic Shortwave   132
Africa   133
The Near and Middle East   139
The Indian Subcontinent   139
The Far East and Asia   140
Australia and the Pacific   143
Canada   145
Latin America   146

8 UTILITY STATIONS   **152**

Longwave Beacons and Other Stations   153
Maritime Communications   157
Fixed Stations   158
Time and Frequency Stations   161
Aeronautical Communications   162

9 OTHER RADIO ACTIVITIES   **168**

Broadcast Band DXing   168
FM and TV DXing   177
Amateur "Ham" Radio   180

10 UNUSUAL, ILLEGAL, AND MYSTERIOUS RADIO ACTIVITY   **189**

Clandestine Broadcasting   189
Pirate Radio   197
"Numbers" Stations   202
Miscellaneous Unusual Signals   207
"Official" Interest in the Shortwave Hobby   210

11 THE HOBBY OF SHORTWAVE LISTENING   **214**

Reporting Reception and QSLs   215
Record Keeping and Awards   222
SWL "Call Signs" and Identifiers   223
SWL Clubs   224

APPENDIX   **229**

Call Sign Allocations of the World   229
International Phonetic Alphabet   234
International Morse Code by Sound Chart   235

Q-Signals Used in Morse Code Communication 236
Abbreviations Used in Morse Code Transmissions 236
Clubs for Shortwave Listeners 237
Commercial Monthly SWL Publications 238
Shortwave Equipment Suppliers 238
Reference Books for the Shortwave Listener 238

INDEX **241**

PREFACE

Listening to shortwave radio may seem like an anachronism in an age when average citizens routinely receive television programs through satellite dishes in their backyards and telephone calls are routed through glass fibers using light. By all logic, shortwave radio should have gone the way of telegraphy over land wires. But the hobby of shortwave listening is bigger—and livelier—today than ever before. Why is this?

Everyone who listens to shortwave radio probably has a different explanation. For me, it's because a shortwave radio is a magic box.

Over twenty years ago I was a junior high school student. I became curious as to why I could hear distant stations on an AM radio at night but not in the daytime. That puzzle led me into library stacks looking for an explanation in the various books on radio. In those books, I found out about some other types of radio signals, known collectively as shortwave, which could be heard all over the world with relative ease. Intrigued, I pestered my parents for a shortwave radio. On my next birthday, my wish came true as I was presented with a simple shortwave receiver manufactured by a now-defunct company known as Hallicrafters.

It was then I discovered the magic inside a shortwave radio.

All I had to do was twist the dial and I could hear programs—in English, no less!—from such countries as the Soviet Union, Great Britain, West Germany, Spain, Australia, and Japan. Another twist of the dial and I heard programs in languages I never heard spoken before, along with strange music I never knew existed. I could hear ham radio operators talking to each other from all over the country and world. I found many stations sending the dots and dashes of the

Morse code; since I didn't know Morse code, they remained a mystery to me. I managed to hear people placing telephone calls from ships at sea along with aircraft in contact with airports. And since this was before the era of communications satellites, I also listened in on many telephone conversations from the United States to Europe.

That shortwave radio became a window on the world for me. Without intending to, I learned much about world politics, customs, cultures, and lifestyles. I found I could automatically name the capital of any country one might mention. I started picking up bits and pieces of various foreign languages and had an endless supply of practice material when I later studied French and Spanish in school. My desire to know more about how my shortwave radio worked pushed me into obtaining my own ham radio license and later into a career as a writer and editor of books on electronics and computing.

Much has changed since I first listened to shortwave radio. There's more to hear today than ever before. Shortwave radios are easier to use, more compact, and a receiver costing less than $200 today can run rings around a receiver that cost over $1000 twenty years ago. In fact, only one thing has remained constant.

The box is still magic.

Harry L. Helms

1

WHAT IS SHORTWAVE LISTENING?

In many ways, the terms *shortwave listening* and *shortwave listener* are misnomers. A person who listens to the news from London over the British Broadcasting Corp. (BBC) is a shortwave listener. Yet those who try to hear stations from Great Britain on the standard AM broadcasting band may also refer to themselves as shortwave listeners (abbreviated as "SWLs," with the singular being "SWL" and shortwave listening known as "SWLing"). SWLs are also found tuning for distant stations on the FM broadcasting band and TV channels or even prowling the frequencies *below* the standard AM broadcast band (the so-called "longwave" band). So what's the difference between an SWL and a normal radio listener or TV viewer? The answer seems to be that SWLs, regardless of the particular wavelengths they listen to, are seeking something out of the ordinary. SWLs want to eavesdrop on signals that the general public doesn't normally receive. Sometimes the public doesn't tune such signals because the equipment required is not widely used (as is the case with most shortwave radios). Or the signals in question might require special knowledge and skills to catch (such as signals from FM and TV stations located several hundreds or even thousands of miles away). SWLs look at the radio and TV spectrum and actively try to find stations that the majority of people aren't aware they can hear.

SWLs are a diverse group. Many specialize in trying to hear "DX" signals. "DX" comes from the radiotelegraph abbreviation for "distance," and refers to stations which are heard only rarely or with extreme difficulty. Other SWLs are content to listen to a certain group of high-power, easily heard broadcasters such as Radio Moscow, the BBC, Deutsche Welle ("Voice of Germany"), and Radio Japan. In publications for shortwave enthusiasts, one often sees the term "DXer" used for the former group and "SWL" for the latter. But the terms are often used interchangeably and it is common for a listener to be a DXer some of the time and a SWL the rest. Most DXers have a few major stations they enjoy listening to on a regular basis and most SWLs enjoy occasionally trying to "dig out" a weak, distant (DX) signal.

The term for a shortwave hobbyist most familiar to the general public is "ham." The formal term is *amateur radio operator*. This refers to someone who *transmits* as well as listens on shortwave and engages in two-way contacts with other amateur radio operators. Such persons must hold a license issued by the appropriate authority of the country in which they are located. (In the United States, the agency is the Federal Communications Commission.) Licenses are issued upon the passing of a written test on radio theory and practice as well as a Morse code receiving test. Many hams got interested in amateur radio through SWLing (as was the case with your author) and still actively engage in SWLing. The public and the media frequently lump all radio hobbyists—SWLs, hams, even citizens band (CB) operators—together as "hams." But the term refers only to those properly licensed to transmit on shortwave.

(By the way, no license or other official sanction is required to own or use a shortwave (abbreviated SW) *receiver*. You can set up a SW radio and antenna and listen as much as you like to anything you wish. However, there are some restrictions on divulging to others certain types of messages you may hear which are not intended for anyone else but a specific recipient. These restrictions will be discussed later.)

Regardless of the label used for a participant, one of the beauties of SWLing is that the hobby becomes whatever the participant wants it to be. Some DXers try to hear signals from as many different nations and stations as possible, and engage in what is, at times, a ferocious competition with other DXers to hear rare or unusual stations. (For many of them, hearing the station is only half the battle; they then try to get a card or letter from the station confirming that they indeed did hear it.) Some SWLs seemingly find DXing a bit vulgar, and content themselves with listening attentively to major international shortwave broadcasters and sending regular letters to such stations commenting on the programs they hear. There are even some monthly publications devoted to shortwave programming; readers discuss favorite or least favorite stations and their programs in a manner (and often at a level) remarkably similar to TV soap opera fan magazines. Others like to listen to communications from barges cruising the Ohio River, to military aircraft aloft, to communications from foreign embassies in Washington, or to illegally operated "pirate" radio broadcasters scattered throughout the United States. Sometimes it seems the only thing SWLs and DXers of different persuasions seem to have in common is a certain ineffable pleasure all seem to find in their activities.

So just what *is* shortwave listening? It's whatever you want it to be, and what you make it. Let's look at some of the types of signals that a SWL or DXer might be interested in receiving.

SHORTWAVE BROADCASTERS

The first stations that attract the attention of a new shortwave radio owner are usually broadcast stations. Such stations are intended for reception by anyone with a SW receiver. Shortwave broadcasters are grouped into two broad categories. *International* broadcasters direct their programming to listeners outside of the country from which the signals are transmitted. Such programs are usually in the language(s) of the intended target countries. For example, Radio Moscow broadcasts extensively to the United States and Canada in English but uses Spanish for most of its programs to Latin America. The Voice of America uses over forty languages for its broadcasts. Most international broadcasters are funded and operated by governments; for example, the Voice of America is operated by the United States Information Agency and receives all funding through Congressional appropriations. In some countries, international broadcasting is under the auspices of a public or statutory corporation; Great Britain's BBC is the best-known such example.

One thing you'll notice is that international broadcasters almost never use call letters (such as WNBC or KFRC). Instead, names or slogans are used to identify such stations.

Government-operated and -financed international broadcasters tend to be little more than public relations outlets for the country. Don't expect startling insights or critical analysis of the sponsoring country. Some stations are more free of control by the funding government than others (the BBC is especially fortunate in this regard), but the overall impression is that the overwhelming majority of state-operated international broadcasters are directing their programs more to the funding authorities in their nation than to listeners in other countries. Program content usually reflects what the funding authorities feel is important for overseas listeners to know about the country rather than what overseas listeners might be interested in. An idealized image of the country is often portrayed in broadcasts, and little effort is made to understand the intended foreign audience. The result is that many programs are simply tedious recitations of agricultural and industrial output statistics interspersed with features on national issues which have little meaning, and less interest, to overseas listeners. (An example is the Voice of Turkey program "Last Week's News In Turkey," which is part of their English broadcasts to North America.) The Voice of America has been criticized repeatedly by American experts on the Soviet Union for failing to understand the audience for its Russian-language broadcasts.

Even dull programming can be useful and informative to the discerning, politically aware listener. Radio Moscow is a case in point. The first reaction upon listening to Radio Moscow might be a mixture of amazement and amusement at the content; those responsible for their programs seem unaware that listeners in the United States have sources of information other than Radio Moscow and can easily spot misleading or false statements. (Radio Moscow often comments on conditions in the United States during its broadcasts to the United States, as if listeners have no other way to find out!) But Radio Moscow's broadcasts do carefully follow the themes and positions being taken in all Soviet media; listening to Radio Moscow is a handy way of understanding how the world is presented to typical Soviet citizens.

Sometimes what is *not* said during international broadcasts is more illuminating than what is said. During the final months of the reign of the Shah of Iran, the English broadcasts from the Voice of Iran took the attitude that nothing was really wrong; there is some noise and confusion, the broadcasts went, but nothing to be concerned about. The Voice of Iran took this line up until the departure of the Shah. Then came a dramatic flip-flop as the Voice of Iran became the Voice of the Islamic Republic of Iran, and invective poured out against the United States. A similar situation took place in Uganda during the final days of Idi Amin. English shortwave broadcasts from Kampala repeatedly warned opposition forces to surrender or face certain death. This pretense continued even as rebel forces advanced on the capital and Amin himself made plans to flee. Listening to broadcasts in such circumstances has a certain surreal quality!

Some international broadcasters are funded and operated by religious organizations. These stations are frankly evangelical in orientation; their purpose is less to entertain and inform than to convert. One example is Trans World Radio, which operates shortwave stations in such locations as Monte Carlo, Swaziland, the Netherlands Antilles, Cyprus, and Sri Lanka. (Interestingly, religious broadcasters are currently entirely Christian. There are no proselytizing broadcasters for other religions such as Islam.)

Finally, there is a small but growing number of private, commercial international broadcasters in operation. One is Radio Trans Europe in Sines, Portugal. You can't listen to programs from Radio Trans Europe itself, since it doesn't produce its own programming. Instead, the station sells air time to broadcasters and organizations who wish to broadcast on shortwave for Europe but who do not (or cannot) establish their own station in Europe. Thus, Radio Trans Europe serves as a relay for such stations as Radio Japan and Radio Canada International, as well as broadcasting programs produced by private organizations like Adventist World Radio.

Private commercial shortwave broadcasters have also sprung up in the United States during recent years. One such station is WRNO, which broadcasts from New Orleans, Louisiana. WRNO uses a format of rock and roll music and lets overseas listeners get a taste of what American domestic radio is like. A similar format—but entirely in Japanese—was tried by KYOI. KYOI is located in the American commonwealth territory of the Mariana Islands in the Pacific; its programs are intended for teenagers in Japan.

Almost all international broadcasters seek contact with and letters from listeners. Sometimes this is out of a genuine desire for opinions and comments from listeners. (Several broadcasters operated by major European nations fall into this category.) Other broadcasters seek listener mail to demonstrate that their programs are indeed being listened to in the intended target countries. More typically, listener contact is solicited so that the aims of the station's funding source can be furthered. As an example, a letter to a religious international broadcaster requesting a program schedule will produce a reply envelope containing several religious tracts and pamphlets in addition to the desired program schedule. A single letter to a government-operated broadcaster can (and often does) place the listener's name and address on the station's mailing list for several years to come. Regular program schedules may be accompanied by tourist information. Sometimes a "hard sell" is employed; Czechoslovakia's Radio Prague once sent, during the Carter administration, a postcard opposing development of the neutron bomb to American listeners on its mailing list. The postcard was preaddressed to President Carter at the White House and required listeners only to sign it, apply a stamp, and mail it.

Why would anyone write an international broadcaster? The prime motivation for most listeners is to obtain a *QSL* or *verification card* from a station. "QSL" is a radiotelegraph code for "I confirm," and the intent of a QSL or verification card is to confirm that the listener indeed heard the station. Listeners write the station a letter, known as a *reception report*, giving information about their reception (time, date, frequency, signal quality) along with enough details about the programs they heard (program titles, announcer names, music played,

Figure 1.1 A "soft sell" approach to promoting South Africa is apparent in this card Radio RSA sends out with QSLs.

and the like) to prove to the station that they actually heard the station. If a listener's report matches the station's records for the time, date, and frequency in question, the station sends the listener a QSL. (The process is not actually so smooth, as will be discussed later in this book.)

Many listeners try to collect QSLs from as many different countries and stations as possible. To keep listeners sending in additional letters after they receive a QSL, many stations change the design of their QSL cards several times each year or offer a series of cards that can be obtained only by sending in a number of reception reports within a specified period of time. Other QSL cards are issued for specific events, such as station anniversaries or the introduction of new broadcasting facilities. Sometimes special QSL cards are sent out for reports on new transmitter sites, when listener reports are especially valuable. Many stations take great pride in the design of their QSLs and produce colorful, artistic cards.

Other souvenirs can be collected from stations. Some stations send pennants to listeners. This practice apparently originated among Latin American stations broadcasting to audiences within their own country but soon spread to major

Figure 1.2 Radio Budapest in Hungary sends out QSLs similar to this.

international broadcasters. Some stations send pennants upon request, while others require a certain number of reports for a pennant. Some stations, such as Radio Moscow, apparently send pennants whenever the mood strikes them. In previous years, most pennants were made of cloth but more recently paper and plastic have been used. Some SWLs have managed to accumulate several hundred different pennants from various stations. Unfortunately, many stations have been forced to curtail sending pennants due to financial considerations.

Figure 1.3 Pennant sent out by Radio RSA in South Africa.

Many stations use postage stamps rather than a meter on envelopes mailed to listeners. Some make it a point to use the latest issues and commemorative stamps. Stations in Eastern Bloc countries send out New Year's cards to listeners and even sponsor listener's "clubs," which offer certificates and "diplomas" to those who send in a specified number of reports.

For years, collecting QSLs from different countries and stations was the cornerstone of the SWL hobby. SWL clubs often featured numerical ranking of members by the number of different countries and stations they had received QSLs from. Accumulation of QSLs from a large number of different countries and stations was viewed as testimony to the listener's skill and experience in receiving signals. Lately, however, some SWLs and personnel at international broadcasters have been critical of the entire practice of "QSLing" and collecting similar souvenirs from stations. Personnel at a few stations have complained that sending out QSLs and other items is an expensive waste of money and staff which could be better utilized producing better, more interesting programs. (This is an argument which implies that creativity and imagination are commodities that can be purchased!) Some stations have even gone so far as to stop issuing QSLs altogether, or have replaced them with "listener cards." Such listener cards look like QSLs but make no pretense of confirming that a listener indeed heard the station; they merely thank the listener for writing. Such station actions have found support among some vocal SWLs who feel that stations should be listened to for program content alone and not to get a QSL.

Yet QSLs seem destined to remain an important part of the experience of shortwave listening. Many listeners still enjoy collecting colorful cards from stations they have heard, and such QSLs can develop considerable goodwill toward the station. Perhaps the most important argument is that stations have yet to devise a more effective means of generating listenership and listener mail than the humble QSL card. It remains a powerful positive reinforcer for desired behavior—listening to the station. (Many listeners collect QSLs from stations other than international broadcasters; this will be covered later.)

Far more interesting to listen to in many cases are the *domestic* shortwave stations of various countries. These are stations intended primarily for reception within the country in which they are located (and sometimes surrounding countries, particularly in Europe). Many countries have sparsely populated areas which can be economically reached by shortwave, and programs (often relays of domestic AM or FM stations) are directed toward residents of such areas. Nations using domestic shortwave broadcasting are not always part of the Third World; Canada, Australia, West Germany, and the Soviet Union have domestic shortwave broadcasters. (Like international broadcasters, most domestic broadcasters identify using names or slogans instead of call letters.)

Domestic shortwave broadcasters offer some of the most fascinating listening you'll find on shortwave radio. You won't hear the loud, booming signals international broacasters use, but you won't find the carefully tailored, sanitized programming either. Programs are in the local language(s) of the countries involved; in addition to Spanish, French, German, or Russian, you'll hear Swahili, Sesotho, Hausa, Pulai, and even Tahitian. You don't have to understand these languages to enjoy listening to domestic broadcasters—wait until you hear the heavy breathing on a Spanish language "radionovella" (soap opera) and the hysterical announcing style in Portuguese of a soccer match from Brazil. And music (even vocals) can be enjoyed without translation. The music you can hear is often totally unlike anything you've heard before (or could hear anywhere else). Several listeners have compiled libraries of music taped from domestic shortwave broadcasters. Domestic broadcasters give something of the texture of life in a country, since you and citizens of that nation are simultaneously listening to the same broadcast. You may be surprised at the impact—or lack of impact—of American culture upon a particular country. (Your author will never forget hearing Gene Autry records being played by a station in Uganda.) And you can hear where elements of American culture, particularly music, had their origins.

DXers find domestic shortwave stations to be among their favorite targets, since they are more challenging to receive. Many international broadcasters use transmitters rated at 250,000 to 500,000 watts (or, as it is usually expressed, 250 to 500 kilowatts, abbreviated KW) of power. These transmitters are connected to efficient, directional antenna systems giving *effective* transmitter powers of over one million watts (one megawatt). Hearing such stations is no problem; in fact, they're difficult to *avoid*. By contrast, domestic broadcasters use but a fraction of the transmitter power and normally use simple antenna systems which don't boost the apparent power of the signal. It's rare to find a domestic broadcaster running even as much as 50,000 watts (50 KW) of power and 1 to 5 KW are more typical power levels. Moreover, almost all international broadcasters try to schedule their programs when reception in the intended target area would be best. Domestic broadcasters operate according to the needs of their home population, and the hours when they are on the air are often not the best ones for reception in North America. Thus, listeners in the United States always find it much easier to receive a transmission from Radio Japan or Radio Beijing than Radio Mil, a domestic broadcaster in Mexico that uses only 250 watts of power.

Fortunately, many domestic stations operate in frequency ranges where few, if any, international stations are found. Three special broadcasting bands have been established solely for stations located between the Tropic of Capricorn and the Tropic of Cancer. These bands were set up because static on the AM broadcasting band is often so heavy in the tropics that reception of AM stations outside the immediate vicinity of the stations is difficult or impossible. These so-called "tropical bands" are favorites of SWLs and DXers worldwide seeking local color or rarely heard stations. Other domestic stations can be heard better when international broadcasters are not usually transmitting to one's listening area. In North America, this is generally the period between local midnight and sunrise.

Numerous domestic broadcasters from Central and South America can be heard throughout the evening and night in North America. Several stations from Africa can also be heard in North America. Perhaps the most exotic listening comes from stations located in Pacific and Asian nations such as Indonesia. Several DXers in the United States devote most of their listening time and effort toward hearing and QSLing stations located in Indonesia. Brazil is another favorite target country.

AMATEUR "HAM" RADIO

Throughout the shortwave spectrum are bands set aside for use by ham radio operators. Hams can be heard communicating with other hams by means of voice, Morse code, and specialized methods such as radioteletype and slow-scan (still picture) television. Hams even communicate through satellites designed and built entirely by hams; these satellites have been launched by American, Soviet, and European rockets. Other hams have recently begun communicating through their personal computers by a method known as *packet radio*. Segments of the ham bands have been allocated to various methods of communication (voice, Morse code, and so on) by informal agreement or by law.

As mentioned before, hams can transmit on shortwave because they have a license issued by the government of the country in which they are located. Such licenses are issued when a prospective amateur radio operator passes an examination on radio theory and (usually) receiving Morse code. Hams are issued unique call signs to identify their stations when they're on the air, and hams become asssociated with and referred to by their call signs. Ham call signs usually consists of one or two letters, a digit, and then one to three additional letters. The letter or letters preceding the digit indicate which nation licensed the station; the alphabet is divided up for this purpose by international agreement. A call sign beginning with W to WZ, K to KZ, N to NZ, or AA through AL always belongs to a station licensed by the United States. Similarly, a call sign beginning with G to GZ belongs to a station licensed by Great Britain. A list of these call sign allocations is in the appendix of this book. (Other types of stations also use this system; this is why broadcasting stations in the United States have call letters beginning with K or W.)

Conversations on the ham bands tend to be among the more interesting or inane you'll ever hear; this is known as "ragchewing" among hams. Many hams like to contact as many different countries as possible, and swap QSL cards with those stations to prove the contact took place (this is also known as DXing). Similarly, some amateurs try to contact all U.S. states or counties. Other hams enjoy sending messages to other cities or states for members of the public; this is known as "traffic handling." Some hams enjoy participating in the numerous on-the-air contests held. The object in such contests is usually to contact as many hams as possible in different states, countries, or other areas.

Some SWLs get so fascinated at listening to hams that they eventually obtain their own amateur license. A few SWLs specialize entirely in listening to hams; they send reports to hams and collect QSLs from hams just as avidly as other SWLs do with broadcast stations.

UTILITY STATIONS

Most stations you can hear on shortwave are not broadcasters. They are *utility* stations. As the name implies, these stations do some type of work. They are not generally intended for general reception, although they can often be heard with little trouble.

Like hams, utilities use voice and Morse code transmission. Other utility stations use *radioteletype*, which produces printed output of messages. Another method is *facsimile*, which is used to transmit weather maps to seagoing ships. Still other utility stations use even more esoteric modes of communication.

There are four major types of utility stations. *Fixed* stations operate from a specific land location and are used mainly to communicate with other fixed stations. *Land mobile* stations also operate from land, but can operate from different locations or while in motion from place to place. *Maritime* stations operate from ships or are land stations used exclusively to communicate with ships. *Aeronautical* stations operate from aircraft or transmit to airplanes in flight.

Some utility stations are operated commercially by such companies as Great Britain's Cable & Wireless, Ltd. Such companies derive revenue by charging for the messages they handle, much like a telephone company. A few companies even operate utility station networks devoted solely to handling messages between units of the company. But the vast majority of utility stations are owned and operated by various governments, with these stations falling into the broad classifications of "civilian" and "military."

Government civilian utility stations are often used to facilitate transportation of some sort, such as air or sea travel. You can listen to aviation weather broadcasts from airports to aircraft aloft along with transmissions to and from aircraft flying international routes. Ships at sea use shortwave radio to communicate with seaports and other ships. Many United States government agencies use shortwave radio to provide communications during emergencies and to back up existing telephone and telex systems. Another significant use of shortwave by the U.S. government is for law enforcement; the Federal Bureau of Investigation, Drug Enforcement Administration, and Customs Service all maintain utility stations that SWLs can eavesdrop on.

Even communications associated with the President and Vice President of the United States can be heard. Many SWLs have been able to hear phone calls placed by the President while aboard Air Force One. Most sensitive communications are scrambled, but sometimes interesting conversations are transmitted "in the clear." An example took place in 1985, when U.S. jets intercepted an Egyptian airliner transporting terrorists and diverted the plane to Sicily. Several SWLs managed to hear "Rawhide" (the code name used for President Reagan) discuss plans for the operation with Secretary of Defense Casper Weinberger as Air Force One flew back to Washington. Such security lapses are rare but keep a small yet enthusiastic group of listeners glued to the frequencies used by Air Force One.

International organizations maintain utility stations and networks. The International Red Cross is one, as is the International Police Organization (Interpol). SWLs can listen in on their activities. Diplomatic services of various nations make use of shortwave facilities, nominally to keep in touch with their home governments. It is possible for SWLs to listen to transmissions from United States embassies in London and Tokyo or to eavesdrop on signals from the Soviet and Polish embassies in Washington.

A useful type of utility station is the *standard time and frequency* station. These stations are operated by various governments on precisely maintained radio frequencies; they also transmit highly accurate time signals (usually obtained from atomic clocks). In the United States, the National Bureau of Standards operates two such stations, WWV in Colorado and WWVH in Hawaii. Both stations can be easily heard throughout North America with their distinctive voice announcements of the time each minute. Other such stations are scattered around the world.

Military forces of all countries make extensive use of shortwave. Like their civilian counterparts, militaries use shortwave to keep in touch with aircraft and ships. In addition, shortwave is also used for communication among separated

Figure 1.4 Colorful QSL from standard time and frequency station WWVH in Hawaii.

land forces. The various branches of the United States military may compromise the largest number of shortwave stations in the world. American military stations range from Coast Guard stations rendering assistance to ships in distress to coded messages transmitted to Strategic Air Command bombers in flight. (Surprisingly, many U.S. military stations will send QSLs in response to reception reports.) The military forces of other countries can also be heard on shortwave, and it is not uncommon for listeners in North America to hear the military forces of Latin American nations (including Cuba). Listeners in Europe often run across signals from Warsaw Pact military forces, including the Soviet Union. Identifying which (or what) military station you're hearing can be a challenge, since so-called "tactical" call signs ("Thunderchief," and the like) are often used along with sophisticated speech-scrambling and encoding methods. But many listeners find such problems to be part of the fun of tuning military stations, as well as other utilities.

AM BROADCAST BAND DXING

As mentioned in the preface, your author's interest in radio was triggered by the fact that stations on the standard AM broadcast band could be heard from cities over one thousand miles distant at night but not in the daytime. You can observe this fact for yourself. Tune across the AM dial at noon your local time. Repeat that at midnight. You'll notice that you'll hear more stations at night, with many of them badly interfering with each other, and that some stations will be from cities hundreds or even thousands of miles away.

Several DXers devote all their listening time and effort to the AM broadcast band. Some of them have managed some impressive results, such as hearing over 100 different countries on AM. With proper conditions, equipment, and listening skill, they have accomplished remarkable feats of reception. Australia has been heard on the East Coast of the United States on AM, for example, and Europe has been heard on the West Coast. AM DXers on the East Coast routinely hear stations in Europe and Africa, while West Coast listeners can often catch stations in Asia and the Pacific. Both coasts have opportunities to hear Central and South America.

Other AM DXers specialize in receiving stations from the United States and Canada. Their goal is to hear (and usually to get a QSL from) at least one station in each state and province; most also try to hear and QSL as many different stations as they can.

As a general rule, it is more difficult to hear an AM broadcast station than it is a shortwave broadcast station over the same distance. Reception conditions (which we'll discuss later) play a major role in AM DX reception; conditions necessary for outstanding reception (such as reception of Australia on the East Coast) may be present only one or two days per year (or may not be present at all during years of high solar activity). Moreover, better receiving equipment and antennas are necessary for AM reception over distances comparable to shortwave, and the level of interference is usually much greater. This is not to imply that AM DXing is superior to SW DXing; it just means they are different. (SW DXing is often just as demanding, and it is possible to hear stations on shortwave frequencies at distances and power levels which are impossible on AM.)

But you don't need exceptional conditions or equipment to get started in AM DXing. Any radio you happen to have—portable, stereo receiver, clock radio, and the like—is capable of pulling in distant stations at night. Several of the skills developed in AM DXing are useful for SWLing, making the AM band a good (and inexpensive) place to start in the hobby of SWLing.

FM BROADCAST AND TV DXING

When channels were allocated for television broadcasting and FM radio, the frequencies were assigned in the belief that they were relatively free of the reception conditions that make possible long distance reception on the AM and shortwave bands. Interference from other stations could significantly degrade picture and sound quality, and the FCC went to considerable lengths (including carefully spacing apart stations using the same channel or frequency) to minimize the possibility of interference.

By and large, the FCC did its job well. You can see this for yourself by tuning across the AM broadcast band, FM broadcast band, and the TV channels at midnight your local time. You'll *normally* find the same stations on FM and TV channels you can hear at noon; reception will be clear with little interference (if any) from stations outside your local area. In contrast, the AM band will be a cacophony of distant and local stations crashing against each other.

The key word in the last paragraph was "normally." On several occasions each year, freak atmospheric conditions make it possible to receive FM and TV stations located hundreds or thousands of miles away. If you have a TV station in your area on channels 2, 3, or 4, you may have seen rolling black bars across the screen during the months of June and July. (The station may even make an announcement that your TV set is not at fault.) Such rolling black bars are caused by distant stations on the same channel trying to break through, and indicate that conditions are right for reception of distant FM and TV stations. FM and TV DXers refer to this as a "band opening" or simply as an "opening."

The conditions which allow for FM and TV DXing are unpredictable (although they are more likely to occur during certain times of the year and during years of high solar activity). This means that you can't plan for FM and TV DX reception the way you can for shortwave; you have to be lucky enough to be listening or viewing when conditions are right. Even when conditions permit DX on FM and TV, the conditions may (and often do) change rapidly and in unexpected ways. For example, some conditions for distant reception may last for a week, while other conditions may last less than an hour. Stations can show up unexpectedly for just a few seconds of reception before disappearing. (This happens when FM and TV signals are reflected off the ionized trails left by meteors entering the Earth's atmosphere.) And the *direction* from which distant TV and FM signals can be heard often changes during an opening. For example, distant TV and FM stations may be heard at first from the west of your listening location. As the opening progresses, stations to the west may abruptly vanish and be replaced by stations to the south of your location. After a few minutes, all distant stations are gone and the FM band and TV channels are back to normal.

Such unpredictability makes FM and TV DXing frustrating to many. However, it means that success depends more upon the ability to recognize unusual reception conditions than it does on equipment. Any ordinary FM radio or TV set can be used. A sophisticated antenna system is not required either; in fact, often a pair of TV "rabbit ears" or the built-in telescoping antenna on an FM portable outperforms a large TV/FM antenna mounted on a rooftop. FM and TV stations also send out QSLs for reception reports from distant listeners and viewers. Unlike shortwave broadcasters and larger AM stations, FM and TV stations are not flooded with reports and often are genuinely pleased to learn that they have been heard or seen many miles away.

CLANDESTINE, PIRATE, AND ILLEGAL RADIO

Not all stations you hear on shortwave operate under international agreements or national laws. Scattered throughout the shortwave spectrum are various stations which operate from hidden locations or in violation of national or international laws. If such stations are extralegal international broadcasters, they are known as *clandestine* stations. If they are aiming for a domestic audience, they are known as *pirates*. The rest, ranging from unlicensed ham-type operations to drug smuggling networks, fall into the broad category of *illegal* stations.

Clandestine broadcasters are almost always operated by a government or with the support of a government. The aim is to influence (and sometimes incite) the population in the target country. An element of deception is normally present in clandestine operations. For example, a clandestine might pretend to be actually operating from within the target country. The true sponsor and purpose of the station are usually concealed, and a fabricated "cover" story may be used instead. The content of clandestine broadcasts is highly political, although the politics may be blended with music and other features to attract listeners. Clandestines appear and vanish according to shifting political currents. For example, Nicaragua was the target of much clandestine activity during the final days of the Somoza regime. (Listeners to one clandestine, Radio Sandino, were instructed on how to fire an automatic rifle and make Molotov cocktails.) After the Sandinistas assumed power, new clandestines opposed to them sprang up.

Pirate broadcasters are privately operated, often as a hobby, in violation of the laws of the country in which they are located. They are generally low-powered and noncommercial; they operate at irregular hours, and the operators are usually young. In the United States and Western Europe, pirate broadcasters typically play much rock music, run satirical skits (some of which are genuinely hilarious), and get obscene at times. The content may be political, but seldom is seriously so. (Usually politics are limited to calls for "free radio," legalizing marijuana, and similar pressing concerns.) The atmosphere is often one of a group of young people having a party that just happens to be on the air for all to hear.

One fascinating aspect of American and Canadian pirate stations is that several have been operated by members of the SWL community who apparently tired of just listening to stations. Some SWLs have received telephone calls from pirate operators announcing they were about to go on the air. Other SWLs and DXers have reported their receptions of various pirate stations to SWL clubs and publications, and have unexpectedly received QSL cards from operators of the stations who also read the same publications or belonged to the same clubs.

Pirate broadcasters are found in other nations as well. In the Soviet Union, such stations are known as "radio hooligans." They broadcast Western rock music and irreverent jokes about the system. In some Third World countries, such as Indonesia and Thailand, some pirates are operated seemingly as a public service for small or isolated communities.

Illegal stations are engaged in two-way communications rather than broadcasting. Converted or modified ham radio equipment is usually employed. One recent "growth area" for illegal radio has been drug smuggling networks. SWLs (and various law enforcement agencies) have listened in on radio networks coordinating drug shipments from ships or planes to delivery points on land. Radio is also used for coordination between drug growing or processing areas and places where shipments are dispatched.

Guerilla groups in Latin America make extensive use of two-way radio. In fact, a major magazine article on Eden Pastora, a leading anti-Sandinista military commander, showed him seated in front of a modern ham radio transceiver. During the final months of the Sandinista campaign against the Somoza government, many SWLs in North America heard cryptic communications in Spanish that were of a military nature. Paramilitary organizations in the United States are also reported to use two-way radio for their activities.

Not all illegal radio is quite so sinister. Many are stations talking to each other in a manner similar to amateur radio operators—but without a license and on frequencies reserved for other stations, such as utilities.

Listeners can also hear mysterious "numbers" stations. These are stations which do little more than transmit groups of numbers. Usually a woman's voice is used to read the numbers, and they are commonly grouped into blocks of four or five digits. Among the languages you'll often hear used are English, Spanish, and German, along with a scattering of others such as Czech and Chinese. It was widely believed that these transmissions are some form of coded messages from intelligence agencies to their operatives in the field, and this has been confirmed by recent spy trials. Both "sides" are apparently using this method. It is known that Spanish language transmissions of five-digit number groups originate from a transmitter site in Cuba. But Spanish language transmission of four-digit groups has been traced to a site in northern Virginia.

THE SHORTWAVE LISTENING HOBBY

Like other hobbyists, shortwave listeners like to keep in touch with other persons who share their interest in SWLing. Since the early days of radio, several clubs and organizations have been organized to allow SWLs to make contact with other listeners, to swap news about what is being heard, and to serve as a forum for opinions on receiving equipment, favorite or least favorite stations, and similar topics. Most of these clubs, all operated on a nonprofit basis by unpaid volunteers, have banded together under an umbrella organization known as the Association of North American Radio Clubs (ANARC). ANARC holds an annual convention for SW enthusiasts, and some of the ANARC member clubs hold their own annual conventions. Clubs run the gamut from the so-called "all band" clubs (those covering shortwave broadcast, utilities, AM/FM/TV DXing, and so on) to those that specialize in a specific area such as shortwave broadcast, AM DXing, and even pirate and clandestine radio.

These clubs serve as the focal points for the shortwave hobby. Although there is no requirement to join a SWL club, most serious SWLs eventually do join one or more clubs according to their interests. Clubs publish bulletins on a monthly (or more frequent) basis, and these bulletins are the best source of current news relating to shortwave activity. Club bulletins contain information on which stations are being heard, changes in frequencies and times used by broadcasters, and details of the latest QSLs received by members. Bulletins often contain member evaluations of new shortwave receivers and accessories, along with articles by more experienced members offering tips on improving shortwave reception, identifying stations broadcasting in foreign languages, and similar subjects.

Local and regional SWL groups have also been formed. These are largely social in nature, and provide a forum for listeners to swap listening experiences, to view QSL collections, and to discuss listening techniques. Some SWLs also form telephone alert groups, in which any listener who notes a band opening or a rare station calls other members and informs them of the news.

Other SWLs rely on commercial publications for information. At the time this book was being written, there was one national magazine devoted to SWLing and related communications topics. There are also several tabloid-style monthly publications devoted to SWLing. A major difference between clubs and such publications is that clubs usually have more recent news and information, since the press deadline for club bulletins is only a few days instead of several weeks.

And clubs consisting of a few hundred members can provide greater contact among listeners than can a publication with many thousands of readers. Yet commercial publications are preferred by some SWLs, since they can run photographs and illustrations which club bulletins avoid because of cost. The editorial quality of commercial publications is also higher, since they are edited and managed by publishing professionals. Finally, commercial publications are generally free of the personality conflicts and club politics which can take up much space in club bulletins. For many listeners, a combination of club memberships and a subscription to one or more commercial publications allows them to keep up to date on the latest news relating to their interests.

Many listeners are content to casually pursue the hobby, but others decide to take SWLing very seriously. Some listeners try to QSL literally every station they hear; if the station doesn't reply to their first reception report, they keep sending additional reports until they get a reply—even if it takes over a dozen reports. Other listeners try to collect as many different QSL card designs as possible, sending the same group of stations a report each month. Specialization is common, with some SWLs devoting all their time and energy to shortwave broadcast, utility, or FM station reception. Perhaps the most serious DXers of all are those trying to hear and QSL as many different countries as possible. They almost invariably have invested considerable money in their receiving equipment and think nothing of crawling out of bed at 3:30 a.m. on a workday so they can try to hear a 500-watt station located in Indonesia or other exotic locale.

The degree of hobby participation is up to the individual. Many, perhaps most, SWLs never join any club; they enjoy tuning on the bands on their own and discovering new stations and frequencies. A few become so wrapped up in the hobby that it becomes the major focus of their life (even to the extent that almost all of their friends are other SWLs). As stated at the beginning of this chapter, SWLing is what you want it to be and what you make it.

TERMINOLOGY AND TIMEKEEPING

Like other specialized interests, SWLing has developed its own terminology and nomenclature. You'll find these terms and expressions used in club bulletins, publications, and by radio stations in their correspondence. We've already introduced some of them (such as SWL, DX, and QSL), and there are others you'll find commonly used. Like "QSL," many originated as radiotelegraph codes and are still used by hams for that purpose. One is "QRM," which stands for *interference*. Another is "QRN," which means *noise* from thunderstorms (sometimes hundreds or thousands of miles away) or from more nearby sources (power lines, neon lights, and the like). The appendix of this book contains a list of other "Q-signals." Finally, you'll often hear shortwave broadcasters wishing listeners "73" on the air or in correspondence. "73" is a traditional radiotelegraph code

Terminology and Timekeeping 19

for "best regards" and is used among stations, SWLs, and hams as a friendly way to end a broadcast, contact, or letter.

Since shortwave radio crosses international boundaries with ease, a worldwide standard of timekeeping is necessary. For this, *coordinated universal time* (abbreviated UTC from the French term for it) is used. You might already be familiar with this system under its old name of *Greenwich mean time* (GMT). This method uses the 0° meridian at Greenwich, England as the standard reference for defining a time for use throughout the world. This means that you'll have to subtract or add some hours from or to UTC, depending on where you're located, to determine the local time equivalent of UTC. The advantage is that a time and day such as "2200 Wednesday UTC" means the same whether you're listening from Tokyo, Paris, or Chicago. All international broadcasters use UTC in their program schedules and generally want time in reception reports specified as UTC. UTC is also the standard time used by hams and utility stations. Moreover, times in SWL club bulletins and commercial publications are almost always GMT (the exceptions tend to be clubs specializing in AM, FM, or TV DX).

If you're listening from North America, you convert UTC to local time by *subtracting* the proper number of hours from UTC. The conversions in Table 1-1 are for major North American time zones:

TABLE 1-1

Time Zone	Subtract from UTC for Local Time
Atlantic Standard	Four hours
Atlantic Daylight	Three hours
Eastern Standard	Five hours
Eastern Daylight	Four hours
Central Standard	Six hours
Central Daylight	Five hours
Mountain Standard	Seven hours
Mountain Daylight	Six hours
Pacific Standard	Eight hours
Pacific Daylight	Seven hours
Alaskan Standard	Nine hours
Alaskan Daylight	Eight hours
Hawaiian Standard	Ten hours

UTC uses the 24-hour or "military" system of time notation. In UTC, midnight is given as 0000. The next hour (or 1:00 a.m. at the Greenwich meridian) is written as 0100. The time fifteen minutes later is expressed as 0115. This system continues with 0200, 0300, 0400, and so on, until UTC afternoon is reached. The next minute following 1259 UTC (or 12:59 p.m. at Greenwich) is 1300 UTC. The time continues with 1400, 1500, 1600, and so on until 2359 UTC is reached; one minute later is 0000 UTC and the start of a new UTC day.

An important point to remember is that when a day is expressed in UTC it refers to the UTC day (or the day at Greenwich). For example, if you want to hear a broadcast scheduled for 0300 Wednesday, and you live in the Eastern Standard time zone, you would listen at 10:00 p.m. *Tuesday* instead of 10:00 p.m. Wednesday. Forgetting to make the necessary day conversion in addition to the time conversion is a common error when using UTC.

2

UNDERSTANDING THE SHORTWAVE SPECTRUM

When you first tune a shortwave radio, you might think the shortwave spectrum is a chaotic jumble of strange signals and noises without any apparent organization. Tune a bit more carefully, and you'll note that certain segments of shortwave are allocated for specific purposes. In this chapter, we'll look at how the frequency range covered by a typical shortwave receiver is divided up. We'll also examine the various methods of transmission, known as *modes of emission*, used by shortwave stations.

Two terms familiar to every SWL are kHz (*kilohertz*) and MHz (*megahertz*). These are the units in which radio frequencies are measured. (These terms replace two older units you may have heard of called kilocycles and megacycles.) A *hertz* (Hz) is one cycle of a radio wave (a cycle is the peak of a radio wave through to the peak of the next wave). One kilohertz is equal to 1000 hertz, and one megahertz equals 1000 kHz. The conversion between these units is as follows:

$$5 \text{ MHz} = 5000 \text{ kHz} = 5{,}000{,}000 \text{ Hz}$$

As a practical matter, SWLs need to concern themselves only with kHz and MHz. With a little practice, you'll soon be able to convert 4895 kHz to 4.895 MHz without any trouble. Both kHz and MHz are used for station frequencies, although kHz is commonly used for frequencies from 0 to 30,000 kHz (30 MHz) and MHz for frequencies above 30 MHz. That convention will be followed in this chapter and throughout the rest of this book.

MODES OF EMISSION

There are several different modes used by radio stations. This section will look at them and explain them in a descriptive, simplified manner. More detailed explanations can be found in such publications as The American Radio Relay League's *Radio Amateur's Handbook*.

One important concept is that of *bandwidth* or *frequency space*. An examination of a listing of shortwave broadcast stations by frequency (such as that found in the *World Radio Television Handbook*) will show that most shortwave broadcast stations operate on frequencies spaced five kHz apart from each other, as in 4800, 4805, 4810, 4815, 4820 kHz, and so on. This is because AM (for *amplitude modulation*, the mode most often used by shortwave broadcasters) occupies at least three kHz, and usually more, of bandwidth. For example, an AM shortwave broadcaster operating on 9800 kHz may actually occupy the frequency range beginning at 9797 kHz and continuing to 9803 kHz. The listed frequency of 9800 kHz is known as the *center frequency* or *carrier frequency*. If you were to tune for the station on a shortwave radio, you would set your receiver's dial to 9800 kHz for best reception. If you tuned slightly off the carrier frequency—to 9798 or 9802 kHz—you could still hear the station, but the audio would probably be distorted and the signal strength would not be as great.

Different modes occupy different amounts of frequency space. For example, a Morse code signal may occupy as little as 100 Hz (0.1 kHz) of frequency space, while the signal of an FM station often occupies more than 30 kHz of frequency space. Obviously, this means many more Morse code (and AM) stations can fit into the frequency space occupied by a single FM station. The differences between modes of emission also affects the receiving equipment needed and the best reception techniques for each mode.

The simplest mode is Morse code, also known as *continuous wave* (abbreviated CW). CW transmission consists of nothing more than simply turning the station transmitter on and off, using a radiotelegraph key or keying device, to produce the dots and dashes of Morse code. (A dash is three times as long as a dot.) As Figure 2-1 shows, a CW signal is very "narrow" and occupies little frequency space. The signal in Figure 2-1 has a center (carrier) frequency of 5000 kHz (5 MHz) and occupies very little space—less than 100 Hz on either side of 5000 kHz.

CW has several advantages in addition to its narrow bandwidth. CW transmitters are simpler and less expensive than those needed for voice or other more complex modes. CW is also the most efficient mode. This is partly due to its narrow bandwidth; 100 watts of transmitter power occupying a bandwidth of 200 Hz is more effective than the same transmitter power spread across a wider bandwidth, such as 6 kHz. Moreover, all transmitter power in CW transmission goes into producing an information-carrying signal. Transmitter power in some other modes go into producing signal components that don't carry useful information.

Modes of Emission

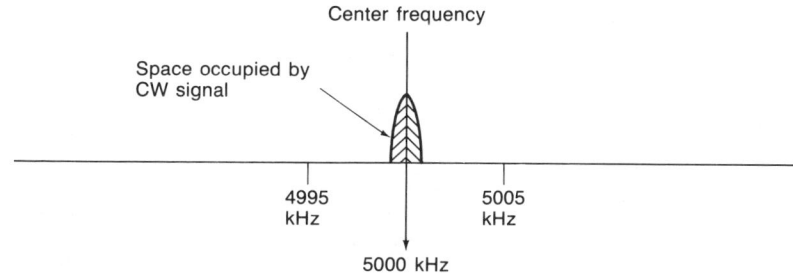

Figure 2.1 A narrow CW signal occupies only a few hundred Hz of frequency space.

The power output level, or amplitude, of a CW transmitter is essentially constant (there is a slight rise and fall of power levels at the beginning and end of each dot or dash). It is possible to vary the output of a transmitter according to sound patterns of voice or music. This is *amplitude modulation* (AM). The amplitude of an AM signal ranges from zero to approximately twice the value of the unmodulated carrier, and the amplitude varies continuously as the voice or music changes.

Figure 2-2 shows an AM signal with a center frequency of 5000 kHz. Note that the AM signal has a set of *sidebands* above and below the carrier frequency. These sidebands contain the information (voice, music, and so on) present in the AM signal, and are each equal in frequency space to the highest audio signal used to vary (modulate) the CW signal. In Figure 2-2, an audio signal of 3000 Hz is used for modulation; the sideband found from 4997 to 5000 kHz is the *lower sideband*, and the sideband at 5000 to 5003 kHz is the *upper sideband*. Both sidebands contain identical information and can be thought of as mirror images of

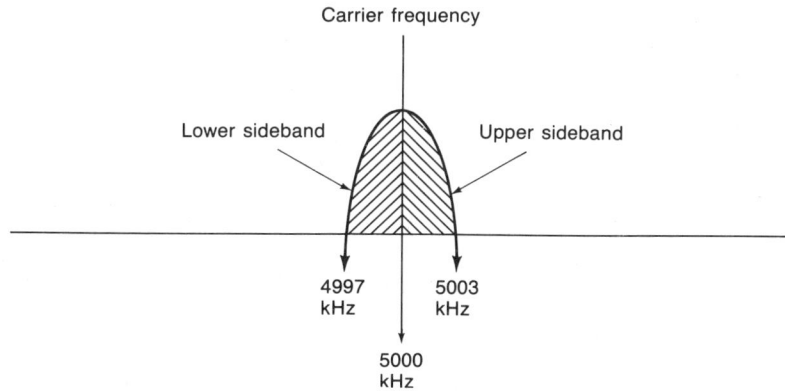

Figure 2.2 An AM signal has "mirror image" sidebands above and below the carrier frequency.

each other. All the useful information in an AM signal is contained in either sideband; the carrier contains no information itself. The transmission of two sidebands is a by-product of the amplitude modulation process, not a deliberately sought condition. The bandwidth of an AM signal is equally to approximately twice the highest audio frequency used; if an audio tone of 5000 Hz were to be transmitted, the bandwidth would be roughly 10 kHz.

AM is easy to produce at the transmitter and simple for a listener to tune. Unfortunately, it is an extremely wasteful mode in terms of transmitter power and bandwidth. Approximately two-thirds of an AM transmitter's power goes into the carrier, which contributes no information. Only one sideband is needed, since both contain the same information. Of the AM signal components in Figure 2-2, all that is really needed to convey the total information available in the signal is either the upper or lower sideband.

That is the rationale behind *single sideband* (SSB) transmission. In SSB, a weak AM signal is generated and the carrier and one sideband are removed. The remaining sideband, either the *upper sideband* (USB) or the *lower sideband* (LSB), is amplified and transmitted by itself. The result is a voice or music signal that far more efficiently uses the available transmitter power than an AM signal; a typical SSB signal has the same efficiency as an AM signal of three to four times the transmitter power. A SSB signal occupies approximately half the frequency space of an AM signal since the other sideband is not used, and thus two SSB signals can fit into bandwidth of one AM signal.

Since SSB sounds terrific, you might wonder why almost all broadcast stations still use AM. The answer lies on the receiving end of a SSB signal. While the carrier in an AM signal doesn't contain any information, it plays an important role when receiving (or detecting) the signal. In a receiver, the sidebands "beat" against the carrier to produce intelligible voice and speech. Tuning a SSB signal with an AM receiver produces a distorted, unintelligible sound that some listeners find similar to Donald Duck's voice.

The solution is to produce a replacement carrier within the receiver itself. This has traditionally been done with a circuit known as a *beat frequency oscillator* (BFO). A BFO is actually an extremely low-power transmitting circuit in the receiver producing a CW signal. The received SSB signal is then beat against the BFO signal so the receiver can then detect the signal in the same manner as an AM signal. (The BFO is also used in receiving Morse code sent via CW; without the BFO, all you hear with a CW signal is a "thumping" sound as the background noise is quieted by the dots and dashes of the code.)

Unfortunately, for many years BFOs were rather crude circuits except on very expensive receivers. BFOs on less expensive receivers were prone to "drift" slightly in frequency, resulting in a signal that was readable only part of the time. Moreover, other receiver qualities are more exacting for SSB reception than AM. (The next chapter will discuss receiver attributes in more detail.)

Happily for SWLs, receiver technology has advanced to the point where almost all contemporary receivers are capable of excellent SSB reception. The

Modes of Emission

availability of satisfactory SSB receiving equipment at a reasonable cost has resulted in a few international broadcasters, such as Radio Sweden, beginning some transmissions in SSB. Other broadcasters can be expected to join them in the years ahead. For years, utility stations and ham operators have been using SSB almost exclusively on frequencies below 30 MHz.

In both AM and SSB transmission, information is transmitted by varying the output of the transmitter in accordance with an audio signal. In both AM and SSB, the frequency of the transmitted signal remains constant. It is also possible to transmit information by keeping signal strength constant but by varying the signal's frequency. This is known as *frequency modulation* (FM). When no audio is being transmitted, the frequency of an FM signal rests at a *center frequency*. As shown in Figure 2-3, the frequency of an FM signal varies above and below the center frequency in accordance with the voice or music being transmitted. The maximum amount by which an FM signal varies above or below the center frequency is known as the *deviation* of the signal. The minimum deviation of an FM signal is usually 5 kHz; since deviation is measured above *and* below the center frequency, the bandwidth of the signal would be 10 kHz. Other common deviations for FM signals include 15 and 30 kHz, meaning that such signals have a bandwidth of 30 or 60 kHz. Obviously, many more AM and SSB signals can fit into the frequency space required for even a single FM signal.

Since FM needs so much frequency space, why is it used? The main reason is that radio noise and electrical interference (QRN) is primarily amplitude modulated. An FM receiver is designed to respond only to FM signals, not AM, and generally "ignores" the AM noise pulses. This means that FM is much less affected by QRN. Also, FM receivers exhibit what is known as the *capture effect*.

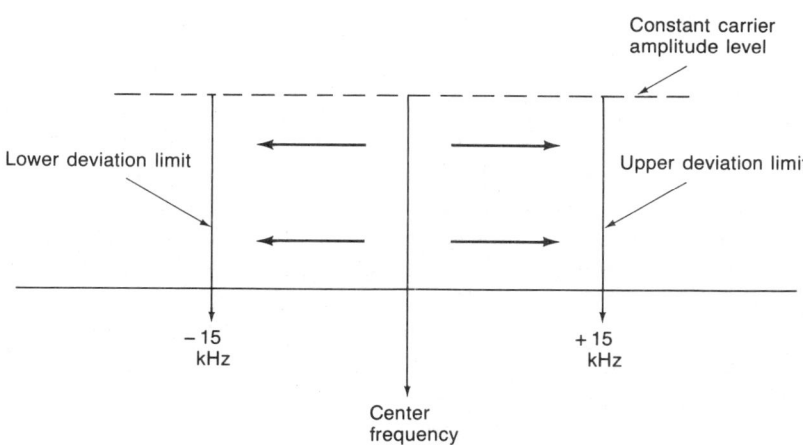

Figure 2.3 The amplitude of an FM signal remains constant; information is transmitted by "swinging" the constant carrier above and below the center ("unmodulated") frequency.

When two or more FM signals are present on the same frequency, an FM receiver tends to respond to only the strongest signal and ignore the others. (This effect is more pronounced if one signal is significantly stronger than the others.) The combination of the capture effect and the general immunity to noise means that FM is capable of much better audio quality, particularly for music, than AM or SSB.

Radioteletype (RTTY) is a mode in which the transmitting station sends messages entered at a keyboard similar to a typewriter or computer terminal. These messages are transmitted by *frequency-shift keying* (FSK). At the receiving end, FSK signals are converted back to written form and are displayed on a video screen or printed out on a printing system similar to that used by personal computers. Some RTTY systems combine the keyboard and printer into a single unit.

When a message is entered at the keyboard of a RTTY transmitting station, the various letters, numerals, and punctuation symbols are converted into a code for transmission. The oldest and still most widely used code is *Baudot*. Recently the *ASCII* (American Standard Code for Information Interchange) code has become popular; ASCII is used for data exchange with personal computers as well. Both Baudot and ASCII are binary codes, meaning that all characters in each can be represented by combinations of two different conditions such as a high or low tone (much as all of the Morse code can be represented by using only dots and dashes). When using FSK, the transmitter switches between two separate radio or audio frequencies to represent the two different conditions making up Baudot or ASCII. The spacing between the two different frequencies is known as the *shift* of a RTTY signal; common values include 170 and 850 Hz. Strictly speaking, FSK is used for those situations where the radio frequency itself is shifted to transmit RTTY. If the radio frequency stays constant and the frequency of an audio tone is being shifted for RTTY, it is termed *audio frequency-shift keying* (AFSK). AFSK is often transmitted using SSB or FM.

RTTY equipment used to be bulky, noisy electromechanical devices. More recently, microprocessor-based RTTY units have come into use along with several peripheral devices to allow personal computers to be used as RTTY terminals. These have allowed SWLs to easily eavesdrop on RTTY signals and have resulted in an explosive growth in RTTY monitoring among SWLs. You will soon learn to recognize a RTTY signal being transmitted by its "tweedling" sound as the signal shifts back and forth; some SWLs even get to the point where they can recognize some characters in Baudot or ASCII.

Two other modes you will run across on shortwave involve the conversion of visual images to a form which can be transmitted by using audio tones. One is *facsimile* (abbreviated FAX), which is primarily used for transmission of weather charts and maps. An image to be transmitted via FAX is scanned by a spot of light. The amount of light reflected back is used to produce a range of audio frequencies which vary according to the reflected light. These audio frequencies are then transmitted. At the receiving end, the audio frequencies control a stylus on a revolving drum carrying paper. The lightness or darkness of the stylus's impression depends upon the audio tone received. The stylus moves from top to bottom on the paper and eventually reproduces the entire map or weather photo. FAX receiving equipment is not commonly available to or used by SWLs. You can recognize a FAX signal by a sound similar to a revolving wheel with a "scratching" noise added.

The other mode by which visual information is transmitted by using audio tones is *slow-scan television* (SSTV). SSTV converts a still image to a series of audio tones for transmission; at the receiving end, these tones are converted to a form in which they can be displayed on a TV set. Both color and black and white pictures may be transmitted. Unfortunately, only still pictures can be transmitted, and it takes several seconds to transmit or receive a SSTV image. Moreover, the equipment necessary is often expensive. SSTV is almost exclusively used in the ham radio bands, although a few international broadcasters have experimented with it. You'll be able to recognize SSTV signals as a rapidly changing sequence of audio tones sounding like the sound effects of a video game; the images will usually be prefaced and followed by voice communications.

There are additional emissions used on shortwave. However, those we've mentioned so far will account for just about everything you'll be able to eavesdrop upon using commonly available gear.

PRIMARY SHORTWAVE FREQUENCY BANDS

The term *shortwave* properly refers only to the 1600–30,000 kHz (1.6–30 MHz) frequency range. However, many so-called "shortwave" receivers actually have broader frequency coverage; a better term for them is *communications* receivers. Frequencies below 540 kHz are known as *longwave*, while the standard AM broadcast band (540 to 1600 kHz) is known as *medium wave*. Frequencies above 30 MHz to approximately 400 MHz are known as *very high frequency* (VHF), and those frequencies above 400 MHz are known as *ultra high frequency* (UHF). All of the frequency ranges are divided into bands for specific purposes.

You'll also soon discover that certain bands are useful for DX reception during the day, and other bands are useful for DX reception at night. A good general rule, however, is that frequencies below 10,000 kHz are best suited for DX reception during evening and night hours; 10,000 to 15,000 kHz is useful for DX during both day and night, and 15,000 to 30,000 kHz are normally useful for DX during daytime and early evening. Frequencies above 30 MHz produce DX reception only during abnormal, exceptional circumstances. Thus, don't expect equal results at the same time on all bands, and plan your listening time accordingly. (The process by which radio signals travel from a transmitting site to listening points is called *propagation* and will be discussed fully in a later chapter.)

You'll note that some frequency ranges are referred to as the "80-meter band," "31-meter band," or some similar designation involving "meter(s)." This is a carry-over from the early days of radio and refers to the *wavelength* of the radio waves in that frequency range. This provides an easy shorthand to refer to a particular frequency range. "80-meters" is easier to say than "3500 to 4000 kHz," but both mean the same.

The following is a summary of the major frequency bands and what you can hear on them. Keep in mind, however, that while these ranges are generally followed there are some variations in different parts of the world. Moreover, there are some "secondary" uses of most bands in addition to the prime use described. Thus, don't be too surprised to find some other types of signals in each band:

Below 150 kHz. Most communications receivers don't tune below 150 kHz, although some military surplus receivers do and some receiving *converters* are available to extend the coverage of receivers into this range. The signals in this range are CW or RTTY, and are usually from military or government stations. One common use for the lower end of this range is for communication with submarines, since those frequencies can penetrate oceans better than any other frequency range. The U.S. Navy's "Omega" submarine navigation system operates on such frequencies as 10.2, 12, and 13.6 kHz. The Soviet Union operates a similar navigation system for its submarine fleet on 15.625 kHz. U.S. Air Force's Strategic Air Command (SAC) also operates stations in this range, since these frequencies are not affected to a significant degree by atmospheric conditions (and therefore would not be rendered useless by the effects of nuclear explosions in the atmosphere). They can be found on 29.5 and 37.2 with coded RTTY transmissions. Other stations you'll find in this range include radio navigation beacons and fixed stations.

150–540 kHz. This is the range most SWLs mean by "longwave." Most activity is CW, although you'll also hear some AM. Many of the stations you'll hear are navigation beacons. Beacons usually do nothing more than continuously repeat their call letters in slow Morse code, and can be heard from 200 to 415 kHz. Unlike other radio stations, longwave beacons do not follow the international allocations for call letters. Instead, the call letters usually suggest the location of the beacon; for example, beacon CLB operates from Carolina Beach, North Carolina, on 216 kHz and MP is located at Montauk Point, New York, and can be heard on 286 kHz. Some of these beacons also include aeronautical weather broadcasts in AM. You can hear the call letters of the beacons being repeated, in Morse code, in the background of the weather broadcasts. The range from 415 to 515 kHz is used for maritime communications in CW. Maritime stations include shipboard stations and shore stations in contact with ships. These maritime stations do use call signs that follow international allocations. Two important frequencies here are 500 and 512 kHz. 500 kHz is an international distress and emergency frequency, and maritime stations around the world monitor this frequency for calls from ships in trouble. 512 kHz is a *calling* frequency;

stations wishing to contact another specific maritime station call it on this frequency and then move to another frequency once contact has been established. 515 to 540 kHz is populated by miscellaneous beacons, some of which can be heard at considerable distances. One is NB, North Bay, Ontario, Canada, which is often heard throughout eastern North America on 530 kHz. Listeners in eastern North America may be able to hear a few broadcasting stations in Europe, which are allowed to operate from 155 to 281 kHz. All frequencies in the 150–540 kHz range are best received during the night, with best DX receiving conditions usually found during the autumn and winter months.

540–1600 kHz. This is the standard AM broadcast band familiar to everyone. In North America and most of South America, AM stations operate on 10-kHz channels beginning at 540 kHz and continuing through to 1600 kHz. In the rest of the world, AM stations operate on channels separated by 9 kHz. A few stations in Central and South America operate on frequencies located between the normal 10 kHz channel spacing. Among these stations are YSS, El Salvador, on 655 kHz and HOF, Panama, on 965 kHz. Like longwave, these medium-wave frequencies are best at night during fall and winter.

1600–1800 kHz. This is an interesting range, formally allocated for radio navigation, that contains a "grab bag" of various radio services. For example, 1610 kHz is widely used by low-powered stations located at airports, national parks, and the like to broadcast information for reception by motorists on their car radios. A handful of radio beacons similar to those found on longwave operate here; the most widely heard is RAB, Rabinal, Guatemala, on 1613 kHz. You can also hear several seismic beacons, used for oil exploration, producing chirping sounds which some listeners describe as "crickets." Cordless telephones are found from 1700 to 1800 kHz, and these can provide hours of *fascinating* listening. At the time this book was being written, there was a possibility that the AM broadcast band in North America could be expanded to include 1600 to 1700 kHz.

1800–2000 kHz. This is known as the 160-meter amateur radio band, and is the lowest frequency range allocated for use by ham radio operators. You'll hear both CW and SSB used here. Most communications you'll hear involve stations located within a few thousands of miles or less of each other, although some hams have managed to contact all continents and 100 different countries on 160-meters.

2000–2850 kHz. The major users of this frequency range are maritime stations, along with some land mobile and fixed stations. Most transmissions are in SSB along with a scattering of RTTY and CW. An important frequency is 2182 kHz, which is a voice channel for emergency and distress maritime communications. Ship-to-ship communications can be found on such frequencies as 2082.5,

2638, and 2782 kHz. You'll also find many maritime weather broadcasts (in USB) from U.S. Coast Guard stations on 2670 kHz. Standard time and frequency stations such as WWV and WWVH can be found on 2500 kHz. In some tropical areas of the world, the 2300 to 2498 kHz range is allocated to domestic broadcasting and is known as the 120-meter broadcasting band. Many of these stations are low-powered and located in such exotic locations as New Guinea, Indonesia, Brazil, and the Falkland Islands. This range will be humming with activity from around your local sunset until your local sunrise.

2850–3150 kHz. This band is allocated for aeronautical mobile stations. One common type is the so-called "VOLMET" stations, which broadcast weather conditions for various aeronautical routes. Airlines also use these bands for communications with their aircraft flying international routes. Almost all aeronautical mobile stations use SSB, primarily USB, with some CW used by a few airlines (primarily Aeroflot, the civilian airline of the Soviet Union). You'll find best reception here from around your local sunset to your local sunrise.

3150–3400 kHz. The prime allocation here is for fixed stations and some mobile stations. For example, the U.S. Department of the Interior operates a net of stations in the U.S.-administered territories of the Pacific on 3385 kHz. The Federal Emergency Management Agency operates over 60 stations scattered throughout the United States on 3341 kHz (along with other frequencies). You can also find standard time and frequency station CHU in Ottawa, Canada, on 3330 kHz during the night. There is another tropical broadcasting band allocation here at 3200 to 3400 kHz; this is known as the 90-meter broadcasting band. Many countries and rare DX can be found on 90-meters.

3400–3500 kHz. This is another aeronautical mobile band. VOLMET stations at Gander, Newfoundland, Canada and at MacArthur Airport on New York's Long Island can be heard alternating on 3485 kHz during the evening.

3500–4000 kHz. This is known as the 80-meter amateur radio band. You'll find CW and RTTY from 3500 to 3750 kHz and voice (usually LSB) from 3750 to 4000 kHz. This is a very popular band and is crowded during the evening hours. There is also a standard time and frequency station on 3810 kHz; it is HS210A in Guayaquil, Ecuador. Listen for the time pulses and Spanish announcements each minute.

4000–4063 kHz. This is a fixed-station band. In the United States, you're likely to hear ham radio operators who are members of the military affiliate radio system (MARS) operating here.

4063–4438 kHz. This is a very active maritime band. Voice communications are in USB, and much CW and RTTY is also used. 4125 kHz is the international SSB ship calling frequency and is busy throughout the hours of darkness.

4438–4650 kHz.   This band is allocated for the fixed and mobile service. One interesting frequency is 4449 kHz, where several U.S. Air Force stations operate in USB.

4650–4750 kHz.   This is another aeronautical band. Of particular interest here are several VOLMET stations located in the Soviet Union.

4750–4995 kHz.   This is the 60-meter tropical broadcasting band, and you'll find it the best band to hear domestic shortwave stations. Stations in Africa begin to fade in around your local sunset until their sign-off at 2300 or 0000 UTC. Among these are the Federal Radio Corporation of Nigeria from Lagos on 4932 kHz, Dakar, Senegal on 4890 kHz, and Contonou, Benin on 4870 kHz. These and other African stations can also be heard again around 0600 UTC when they sign on (listeners in western North America often find this a better reception time). The evening and night hours are dominated by signals from Latin America. Familiar signals in North America include Radio Barquisimeto in Venezuela on 4990 kHz, Radio Nacional de Colombia on 4955 kHz, Radio Colosal in Colombia on 4945 kHz, and Emisora Gran Colombia in Ecuador on 4910 kHz. Stations from the Pacific and Asia can be heard from about 0730 UTC until shortly after your local sunrise. Among the stations you can hear include VLM4, Brisbane, Australia, on 4920 kHz relaying the Australian Broadcasting Corp.'s domestic service and the National Service of Papua New Guinea on 4890 kHz.

4995–5005 kHz.   This range is set aside for standard time and frequency stations worldwide, most operating on 5000 kHz. During the night, listeners in North America will hear WWV, Fort Collins, Colorado, on 5000 kHz. They broadcast an announcement of the time in UTC each minute, using a man's voice. If you listen carefully after 0500 UTC, you will likely hear another station underneath WWV announcing the time each minute using a woman's voice. That will be WWVH in Kauai, Hawaii. During unusual reception conditions, you can hear other time and frequency stations in China, England, India, Japan, and the USSR in this range.

5005–5450 kHz.   This range is primarily used by fixed and land mobile stations, although some tropical broadcasting stations may be found in the lower 60 kHz of the range. You'll find SSB, CW, and RTTY used here throughout the evening and night hours.

5450–5730 kHz.   The first 30 kHz of this range is shared by fixed stations and aeronautical stations; the remainder is exclusively aeronautical worldwide. Most communications will be in USB and involve VOLMET broadcasts and long-range communications by aircraft flying international routes. Aircraft flying the Caribbean can be found on 5520 and 5550 kHz, while 5598 and 5649 kHz are used by airplanes flying north Atlantic routes to and from Europe and North America. As with the other bands mentioned so far, you'll find the most activity during the evening and night hours.

5730–5950 kHz. This is the range assigned to fixed stations, and numerous stations in SSB, CW, and RTTY can be found in this range. The U.S. National Weather Service maintains a network of stations using USB on 5923 kHz. The Department of Energy uses 5948 kHz to coordinate shipments of nuclear materials across the country. And the U.S. Air Force and the National Aeronautics and Space Administration (NASA) use 5810 kHz for USB support communications for space shuttle launches. Try tuning to this frequency whenever a launch is scheduled, particularly if you're located within 2000 miles of Cape Canaveral.

5950–6200 kHz. This is the first frequency range allocated for international broadcasting, and is known as the 49-meter band. You'll find this range packed with AM signals from late afternoon until approximately an hour after your local sunrise. In addition, you may find some signals (although usually weak) during the daytime, especially in winter. During the evening hours in North America, this band is full of powerful signals from European international broadcasters. Some interesting signals can be found amid the powerhouses, however, such as Radio Luxembourg on 6090 kHz. Radio Luxembourg is an English-language pop music station patterned after American "hit music" radio. It lets American listeners hear the latest European hits long before they make it (if ever) to American music charts. Try for it from around your local sunset until a couple of hours later; best reception is often in May and early June. Around 0500 UTC, the European international broadcasters start leaving the air because Americans are going to bed (unless they're SWLs), and sunrise is starting to spread across Europe. The reduced interference on 49-meters then lets you hear several stations in South America which operate all night in Spanish. Around 0800 UTC, stations from the Pacific and Asia start to become audible. China, Indonesia, Australia, the USSR, and the Philippines are among the countries you can hear in the hours before and shortly after your local sunrise.

6200–6525 kHz. This range is allocated exclusively throughout the world for maritime communications, and will be busy during the same hours the 49-meter broadcasting band is in use. As with other maritime bands, you'll hear SSB (usually USB) used along with CW and RTTY. 6218.6 kHz will be in use almost constantly for USB communications; this is used for inland (that is, rivers and the like) maritime messages ("traffic") as well as by the U.S. Coast Guard along the Atlantic and Pacific coasts. Another active USB channel is 6221.6 kHz.

Primary Shortwave Frequency Bands

Figure 2.4 QSL card from Radio Luxembourg.

6525–6765 kHz. This is another aeronautical band populated by VOLMET stations, aircraft aloft, airports, and flight operations centers. The U.S. Air Force (USAF) uses such frequencies as 6670, 6683 (often used by Air Force One), 6712, and 6738 kHz for its USB communications. You may hear messages on these frequencies composed of various words and numbers; these are known as "Sky king" or "Foxtrot" broadcasts and are coded instructions to USAF aircraft aloft.

6765–7000 kHz. This is allocated for fixed stations. One interesting frequency is 6905 kHz, which is where several stations operated by Interpol operate with RTTY. You can also hear KKN50 in CW on 6925 kHz; this station is operated by the U.S. Department of State from a location near Washington, D.C., to communicate with U.S. embassies and consulates in other nations.

7000–7300 kHz. This range is shared by both amateur radio and international broadcasting; it is known as 40-meters. The 7000–7100 kHz range is allocated exclusively to amateur radio throughout the world. 7100–7300 kHz is used for international broadcasting in Europe and Asia, while it is assigned to hams in North and South America. The result is that this may be the most crowded band on shortwave, as tuning it during the evening hours will reveal. For international broadcasting, it is very similar to 49-meters in what you can hear and when. Hams use LSB in addition to CW and RTTY. In the United States, 7150–7300 kHz is used for voice and SSTV, while 7000–7150 kHz is where CW and RTTY are found. During the day, hams use this band for communications over distances of approximately 1000 miles or less.

7300–8195 kHz. This is assigned worldwide for the fixed service, although some international broadcasters can be found in the lower end of the range. You can hear CHU from Canada on 7335 kHz with time announcements in English and French each minute. Interpol stations can be found on 7401 kHz in RTTY. The U.S. Customs Service also uses 7527 kHz in USB to coordinate its activities.

8195–8815 kHz. This is another busy maritime band, with CW used more heavily than USB. 8257 kHz is an international ship calling frequency for USB; 8291.1 and 8294.2 kHz are other common USB frequencies. 8364 kHz is an international CW frequency for emergency and distress signals.

8815–9040 kHz. This is another aeronautical band. 8825 kHz is used for USB traffic by flights crossing the Atlantic. 8870 kHz is used for VOLMET broadcasts by stations in Gander, Newfoundland and Long Island, New York. Air Force One has often been heard on 9018 kHz in USB.

9040–9500 kHz. This is allocated to fixed stations, with many using RTTY. You'll also find a few international broadcasters scattered through this range.

9500–9900 kHz. This is known as the 31-meter international broadcasting band, and may be the most heavily used band for that purpose. It is also a "transitional" band; best reception is generally during evening and night, but some stations can usually be heard on this band during the day (especially in winter). During late afternoons, stations from Europe and Africa "fade in" to audibility. During evenings, almost every major European international broadcaster can be found here. These stations leave the air around 0500 to 0600 UTC and are replaced by stations from South America, the Pacific, and Asia. Stations from these areas can often be heard two or three hours after your local sunrise; Australian stations can often be heard until approximately 1400 UTC on the east coast. One you might want to try for is VLW9, on 9610 kHz, transmitting the Australian Broadcasting Corp.'s domestic service from Wanneroo in the province of Western Australia. For most SWLs in eastern North America, this will be the most distant broadcasting station that can be heard. The upper end of this range is still used by various utility stations, since that portion was assigned to them prior to 1979; the remaining utilities will be leaving in the years ahead.

Primary Shortwave Frequency Bands

9900–9995 kHz. This range is allocated worldwide to fixed stations, most of which use RTTY.

9995–10005 kHz. This is reserved for standard time and frequency stations. Many of the stations found at 4995–5005 kHz can also be found here, with WWV heard around the clock in North America.

10005–10100 kHz. This is an aeronautical band used largely by airplanes aloft. Interesting USB frequencies include 10072 and 10075 kHz, which are used by airline companies to communicate with their aircraft on matters unrelated to navigation and safety (communications on those topics are handled by airports).

10100–10150 kHz. This is the 30-meter amateur radio band, which was allocated to amateurs in 1979. Because of the narrowness of this band, operation here is normally restricted to CW and RTTY.

10150–11175 kHz. This allocation is reserved for fixed stations worldwide. In addition to the types of utility communications mentioned previously, this range has many *broadcast feeders*. A broadcast feeder is used by international broadcasters to relay programs from their studios to overseas transmitter sites. Such feeders are gradually being replaced by satellite relays, but a surprising number are still operated by such broadcasters as the Voice of America, Radio Free Europe, and Radio Moscow. You'll recognize feeders since they carry normal broadcast programs, but SSB is used instead of AM. You'll also find numerous Interpol stations on 10295 kHz in RTTY.

11175–11400 kHz. This is an aeronautical allocation. 11182 kHz is an active USAF channel, and Russian language VOLMET stations in the Soviet Union can be heard on 11279 kHz. Aeroflot flights enroute between Moscow and Havana can be heard on 11312 kHz—but all transmissions are in CW, not SSB.

11400–11650 kHz. This is a fixed-station allocation with many RTTY and facsimile stations found here. A few international broadcasters can be found in the upper end of this range.

11700–11975 kHz. This is the 25-meter international broadcasting band and often has something interesting to hear around the clock. You'll find most of the major international broadcasters of the world using this band, along with several international broadcasters which do not have programs specifically targeted to North America. For listeners in North America, this is a good band to tune during late mornings and early afternoons for stations in Europe, Africa, and the Middle East. During the evening, this band is often used by major international broadcasters for their North American services. Later in the night, stations from the Pacific and Asia become audible and can be heard until shortly after your local sunrise. One favorite of many SWLs is the Radio Television Francaise d'Outre Mer (RFO). station at Papeete, Tahiti, on 11825 kHz. In addition to some authentic "South Seas" music, it may be your only opportunity to hear spoken Tahitian!

11975–12330 kHz. This is assigned to fixed stations worldwide, although some international broadcasters are often found in the lower range. Reception characteristics are very similar to 25-meters. An interesting frequency is 12216 kHz, where the Federal Emergency Management Agency (FEMA) operates several stations using USB. Several broadcast feeders also operate in this band.

12330–13200 kHz. This is a maritime allocation, and is quite busy throughout the day and early evening hours. One very busy USB frequency is 12429.2 kHz; however, most of the signals in this band are CW or RTTY.

13200–13360 kHz. This is an aeronautical band. Active frequencies used by the USAF for USB communications include 13201 13204, and 13241 kHz; the latter channel is often used for "Sky king" broadcasts.

13360–13600 kHz. This is allocated to the fixed service. It also lies at the beginning of the part of the radio spectrum in which best reception takes place during the day and early evening (and, as will be detailed later, during years of high solar activity). Among the interesting stations in this range are those operated by the Bulgarian, Hungarian, and Yugoslav embassies in Washington, D.C.

13600–13800 kHz. This band was reassigned to international broadcasting in 1979 and is currently shared with fixed stations. Over the next few years, you will likely hear fewer utility stations in this range and more international broadcasters.

13800–14000 kHz. This is a fixed-station band. Among the stations found here is an emergency network operated by the International Committee of the Red Cross on 13915 and 13997 kHz in USB. The FCC also uses 13990 kHz for RTTY communications between its monitoring stations.

14000–14350 kHz. This is the 20-meter amateur radio band and is the best ham band for DX communications. The first 100 kHz is reserved for CW and RTTY, and 14100–14350 kHz is used for SSB (normally USB) and SSTV (although U.S. hams can use only 14150–14350 kHz for voice or SSB). This band is most useful (or "open") for DX communications during the daytime and evening, although it is often open around the clock during years of high solar activity.

14350–14990 kHz. This is another band for fixed stations, and you'll find CW, SSB, RTTY, and FAX used here. Some stations you can hear include a network of Australian research bases in the Antarctic using USB on 14415 kHz and an Interpol network for African nations on 14827 kHz in CW.

14990–15010 kHz. Like 4995–5005 and 9995–10005 kHz, this is a standard time and frequency allocation; many of the stations you can hear on 10000 kHz can also be heard here. A good time to listen for a station other than WWV or WWVH is to tune around your local sunrise, when signals from those two stations are most disturbed.

15010–15100 kHz. This is a narrow aeronautical band. A heavily used USAF frequency is 15041 kHz in USB—listen for the various tactical call signs such as "Morphine," "Tomahawk," and "Overbrook." 15048 kHz is another busy USAF channel, and sometimes Air Force One can be heard there. A few international broadcasters intrude upon this band. The BBC can be heard during the daytime on 15070 kHz, while Iran uses 15084 kHz during the day for Farsi and Arabic programs.

15100–15600 kHz. This is the 19-meter international broadcast band, although some fixed stations can be heard in the upper end of the range. This is a heavily used band during the day and evening hours. It is particularly useful for reception of Asian and Pacific stations during the evening during summer months; try for RFO in Tahiti on 15170 kHz. Radio Japan, Radio Beijing, and Radio Australia frequently use this band for their evening transmissions to North America. This frequency range, like the 20-meter ham band and all higher frequencies, offers best reception during years of high solar activity (indicated by a high number of sunspots).

15600–16460 kHz. This is a fixed-station allocation worldwide. A network of USAF stations in USB can be found on 15632 kHz.

16460–17360 kHz. This broad range is shared by fixed and maritime stations. 16523 kHz is an international USB calling frequency for ships. Other active USB ship channels include 16587, 16590, and 16593 kHz. Station ROT, Moscow, USSR, can be heard in CW on 17130 kHz; this station is operated by the Soviet Navy.

17360–17550 kHz. This is an allocation primarily for fixed stations, although Air Force One has been heard on 17385 kHz in USB.

17550–17900 kHz. This is the 16-meter international broadcasting band and is useful for daytime reception from stations located to the east of your location. If evening and night reception on 19-meters is good, try for Pacific and

Asian stations here as well. Some fixed stations can also be found in the lower end of the band.

17900–18030 kHz. This narrow range is allocated for aeronautical use. The USAF uses 17975 kHz for USB communications, and 17925 kHz is reportedly used for emergency USB transmissions during airplane hijackings.

18030–19990 kHz. This range is allocated, with some exceptions, to the fixed service. 18068 to 18168 kHz is scheduled to be reallocated to amateur radio in the near future (it may have happened by the time you read this), while 18780–18900 and 19680–19800 kHz are shared with maritime stations. This band is generally useful only during the daytime and may not be able to support DX communications during years of low solar activity.

19990–20010 kHz. This is an allocation for standard time and frequency stations (WWV is found on 20000 kHz) and space satellite transmissions. This range is often used by the Soviet Union for both its unmannned and manned space programs, with CW used along with encoded telemetry signals. 19994 and 20008 kHz are two commonly used frequencies.

20010–21000 kHz. This broad allocation is exclusively for fixed stations worldwide. 20042 kHz is used for USB communications by the USAF, and 20053 kHz is used by Air Force One. NASA uses 20191 kHz for LSB communications in connection with launch support activities.

21000–21450 kHz. This is the 15-meter amateur radio band, and ranks second only to 20-meters in popularity for DX communications. In fact, during years of high solar activity it can outperform 20-meters for DX. The first 200 kHz are usually occupied by CW and RTTY signals, while the remainder is used for USB and SSTV.

21450–21850 kHz. This is the 13-meter international broadcasting band, and is most useful during the daytime. During years of low solar activity, many broadcasters abandon this range for 16-meters and 19-meters. You'll also find some fixed stations at the upper end of this range.

21850–21870 kHz. This narrow frequency segment is reserved for fixed stations.

21870–22000 kHz. Although assigned for aeronautical stations, several fixed stations also are found in this range.

22000–22855 kHz. This is an active maritime allocation during day-

Primary Shortwave Frequency Bands

time. 22124 kHz is reserved for USB communications between ships, while 22127 kHz is used by the USCG.

22855-23200 kHz. This band is allocated to fixed stations. During years of low solar activity, you will find very little here.

23200-23350 kHz. This is an aeronautical band, with 23337 kHz used for USAF USB traffic.

23350-24890 kHz. This is allocated to fixed stations worldwide. Its usefulness for DX communications depends very heavily on the level of solar activity.

24890-24990 kHz. This range was recently reassigned from fixed stations to amateur radio, and is known as the 12-meter band.

24990-25010 kHz. This range is assigned to standard time and frequency stations, although none operate here at this time.

25010-25550 kHz. This range is assigned to fixed, land mobile, and maritime stations. Many of these stations, particularly in land mobile service, are low-powered units such as those used in taxis, boats, in plants and factories, and the like. However, surprising distances can be covered with low power during years of high solar activity. For example, listeners on the West Coast have reported what seem to be Indonesian and Chinese language taxi dispatching communications here.

25550-25670 kHz. The range is set aside for use by radio astronomy.

25670-26100 kHz. This is the 11-meter broadcasting band. However, it is seldom used even during periods of high solar activity.

26100-28000 kHz. This is set aside for fixed and mobile communications. In the United States and Canada, 26965 to 27405 kHz is used for citizen's band (CB) radio. Illegal two-way communications by "outlaw" CB operators can be found in the 27500-28000 kHz range; many of these will sound similar to ham radio operations, but they are not.

28000-29700 kHz. This is the 10-meter amateur radio band. It is useful for DX communication during years of high sunspot activity and for local communications.

29700-30000 kHz. This range is mainly used for land mobile stations.

3

SELECTING A SHORTWAVE RECEIVER

Selecting a new shortwave radio can be difficult, especially if it's your first receiver. The choices are more numerous than ever before, with price tags running from slightly more than $100.00 to well into the thousands. The specifications can be baffling to shortwave newcomers; just what do "4 kHz @ -6 dB" and "50 Ω input" mean anyway? In this chapter, we'll look at the different types of shortwave radios and the meaning and importance of various receiver specifications.

Sometimes you'll see shortwave receivers described as *communications* receivers. This term usually refers to more advanced receivers capable of receiving various types of signals (AM, CW, SSB, and so on) under difficult reception circumstances. Another term you'll often see is *portable* receiver. As the name indicates, this is a radio which can be carried about easily and operated from batteries. The distinction between these two types has become blurred in recent years. Many portable receivers now offer performance and features formerly available only from communications receivers, while many new communications sets are light and compact enough to be easily transported and can operate from larger batteries such as those found in automobiles.

RECEIVER BASICS

It helps to know a little about how a typical shortwave receiver works. This makes it easier to understand the importance of various receiver specifications and to determine which features are important and which are merely "window dressing."

Receiver Basics

Figure 3.1 An example of the dramatic advances in receiver technology is the Sony ICF-2002 portable; about the size of this book, it offers direct frequency readout and SSB capabilities.

Most receivers today are *superheterodyne*. This means that the received frequency (such as 9500 kHz) is changed to another, fixed frequency (such as 455 kHz or 10.7 MHz) before the radio signal is converted to an audio signal you can hear. Sometimes a received radio signal is converted to two or even three different fixed frequencies before finally being changed to an audio signal; this is known as *double* or *triple conversion*. If only one fixed frequency is used, it is called *single conversion*. The superheterodyne design is used because certain functions, such as amplifying the received signal, are more easily and efficiently accomplished at a single frequency than over a wide range of frequencies.

Figure 3-2 shows a block diagram of a single-conversion superheterodyne receiver suitable for AM, SSB, or CW reception. Radio signals striking the antenna produce weak electric currents in the antenna. These weak currents are amplified in the *radio frequency amplifier* section. The amplified signal is then applied to the *mixer* stage. Note that a signal from the *local oscillator* section is also applied to the mixer. As its name implies, the mixer combines the signals from the radio frequency amplifier and local oscillator to produce a new, fixed signal (such as 455 kHz or 10.7 MHz) known as the *intermediate frequency*. Regardless of the actual frequency the radio is tuned to (9500 kHz, for example), the intermediate frequency remains constant.

The mixer is usually followed by an *intermediate frequency amplifier*. This

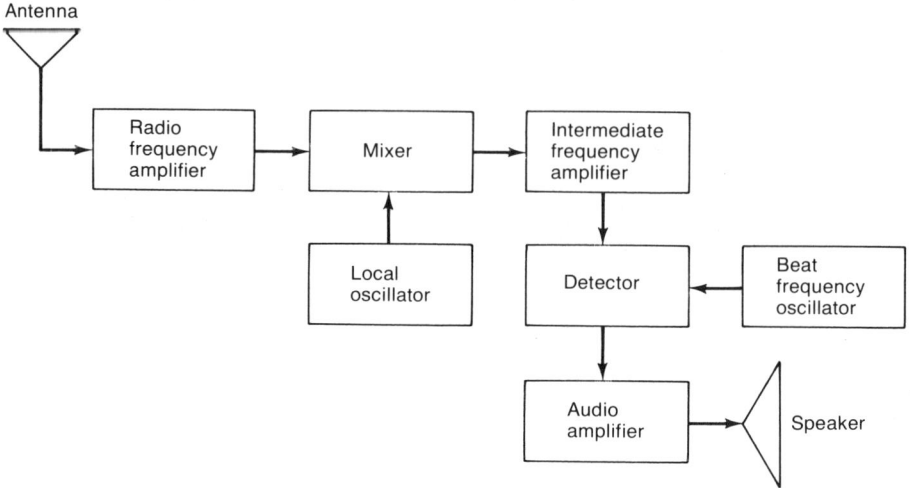

Figure 3.2 Block diagram of a superheterodyne receiver.

stage amplifies the intermediate frequency and feeds it to the *detector*. The detector is the receiver section that converts the intermediate frequency into understandable audio. If the signal received is CW or SSB, the *beat frequency oscillator* supplies a replacement carrier to make the signal intelligible. Finally, the *audio amplifier* stage amplifies the detected signal so you can hear it through a speaker or headphones.

If double or triple conversion designs are used, the different mixer and intermediate frequency amplifiers are usually "chained" together one after another. Double or triple conversion is used, because certain receiver functions and features can be implemented better at one fixed frequency than another. This will be examined later in this chapter.

FREQUENCY COVERAGE

Most shortwave radios are what are known as *all-band* or *all-wave* receivers, meaning that they cover at least 540 to 30000 kHz. Most modern SW receivers also include coverage down to about 150 kHz or so. An all-band or all-wave receiver can cover this entire range without any gaps in its coverage or missing frequencies. Some receivers include additional coverage, such as the FM broadcast band, but a tuning range of at least 540–30000 kHz is adequate for almost all listening interests.

In previous years, many shortwave receivers tuned only certain segments or bands, usually in 500-kHz-wide portions. For example, a receiver could tune 40-meters from 7000–7500 kHz or 6900–7400 kHz, while 31-meters could be tuned from 9500–10000 kHz. A receiver could usually tune a dozen or so 500-kHz-wide segments, with users able to determine which segments they wished to tune by inserting the proper crystal for a desired range into the radio's tuning circuitry. Such receivers achieved great popularity among SWLs and are still used today by many. Among the most popular were the R. L. Drake Company's models R-4B and SPR-4 (now discontinued). Similar receivers covering only the amateur radio bands were commonly used by hams. While no longer manufactured, such receivers are often available used from SWLs, hams, and shortwave equipment dealers. The problem is that crystals for additional tuning ranges are no longer supplied by some manufacturers of the receivers.

Almost all shortwave receivers manufactured today are all-wave in coverage. The only exceptions are a few portables, primarily at the lower end of the price spectrum, which cover only the major international broadcasting bands. Unless price is a very crucial factor in your purchase decision, the best bet is an all-band receiver. Such a radio will let you sample and experience the full range of radio listening activity and won't become obsolete if your interests expand or change.

FREQUENCY READOUT AND TUNING

The biggest advance in SW radio technology in recent years has been in the area of frequency readout. Only a few years ago, finding a specific desired frequency was a bit like shooting in the dark—sometimes the target was found, but more often it was missed altogether. This was because moving the tuning dial as little as one inch could cover over 500 kHz of frequency space; in addition, dial markings were often given only every 500 kHz and even then were usually inaccurate by 100 kHz or more. Some receivers included a fine tuning or "bandspread" dial for the various ham radio bands, but even these were no more finely calibrated than every 10 to 25 kHz.

Today the situation is vastly improved. Most shortwave receivers include *direct frequency readout*, which displays the frequency by using light-emitting diodes (LEDs) or a liquid-crystal display (LCD). Sometimes the desired frequency is reached by turning a dial until the LED or LCD display shows the proper frequency, while in other radios the frequency can be entered by using a keypad similar to that found on a touch dialing telephone. A few radios allow both methods to be used. There is no difference in LED and LCD dials for SWL purposes. Both can display frequencies with equal accuracy. An LED display "glows" when in use and is more readable in darkness or dim light, while an LCD display is visible due to reflected light and can be used only in lit areas or by switching on a light in the receiver. However, an LED display consumes more power than an LCD display. This does not matter when operating the receiver from an AC wall outlet but can be crucial in portable radios or when batteries are used.

Figure 3.3 The Icom R-71A is a high-performance receiver favored by many DXers.

While they are a major advance over previous tuning methods, LED/LCD frequency readouts do involve some factors you should be aware of when comparing receivers. One is the *tuning step* used by the receiver. Many contemporary shortwave receivers use tuning circuits based on digital electronics; they tune frequencies in steps or *increments*. (Such tuning circuits involve *phase-locked loops*, abbreviated PLL.) For example, a radio may tune frequencies in steps of 0.1 kHz. This means that frequencies increase or decrease as the dial is turned by increments of 0.1 kHz (or 100 Hz); if the radio is tuned to 9500 kHz, tuning the dial upward "steps" the receiver through the frequencies of 9500.1, 9500.2, 9500.3, 9500.4, and 9500.5 kHz, and so on. The receiver would be unable to precisely tune such frequencies as 9500.01 or 9500.03 kHz, since it would "step across" them. In practical terms, this is not important since even narrow modes such as CW occupy at least 100 Hz of frequency space, so there is no possibility of missing any signals. Many communications receivers have multiple tuning steps (or *speeds*). This allows listeners to rapidly tune to a desired frequency by using a faster tuning speed (such as a 1-kHz increment) and then to fine-tune a signal by switching to a slower speed such as a 0.1- or 0.01-kHz step.

For most listening, a tuning increment of 0.1 kHz or less will be satisfactory. However, some receivers (mainly less expensive portables) have a tuning increment of 1 kHz or even 5 kHz. (Receivers using a 5-kHz increment have a "fine tuning" control to tune between the 5-kHz steps; however, the receiver frequency readout is not affected by the fine tuning control.) Such radios will do fine for listening to broadcasters (which, as was mentioned earlier, usually operate on frequencies spaced 5 kHz apart). However, they are far less useful for listening to utilities and hams or such modes as CW and RTTY. The *only* sure way to determine the tuning increment(s) available on a receiver is to examine the specifications in the owner's manual or manufacturer's literature for the receiver in question. If the LED/LCD dial includes a decimal point (for a frequency such as 9500.1 kHz), you can be confident it has a tuning increment of at least 0.1

kHz. If it doesn't, it could have a tuning increment of 1 or 5 kHz. If you're not sure, ask the receiver dealer or examine the owner's manual.

One interesting quirk in some receivers is that the *resolution* of the frequency display may be less than the tuning rate. One well-known and popular receiver has a display resolution of 0.1 kHz, meaning it can display such frequencies as 9500.1 kHz. However, it has a slow tuning rate of 0.01 kHz available. This means that the receiver can tune such frequencies as 9500.10, 9500.11, 9500.12, 9500.13, and 9500.14 kHz, yet the frequency display will still read 9500.1 kHz. It will remain so until 9500.20 kHz is tuned, at which point it will read 9500.2 kHz. However, this is seldom a problem in actual use of this receiver and similar radios. Some receivers incorporate a *receiver incremental tuning* (RIT) control, which is a fancy term for a fine tuning control. This allows tuning frequencies in between the various tuning steps, but usually does not change the LED/LCD readout. RIT can be handy for best CW and SSB reception if a receiver tunes in 0.1-kHz steps, since it permits tuning for the most natural-sounding SSB speech or most pleasing Morse code sound. If a receiver can tune in 0.01-kHz increments, RIT is generally not necessary.

Only a few new shortwave receivers are available today without LED or LCD readout, and are mostly portables in the lower price ranges. The extra cost for direct frequency readout is well worth the money. Without an LED/LCD dial, you'll find it extremely difficult, if not impossible, to locate the specific frequencies mentioned in this book and other SWL publications or even to tune to your favorite stations. In the past, many new SWLs abandoned the hobby after a few months because of the frustration of trying to find desired frequencies. Direct frequency readout dials make finding a frequency as easy as dialing a telephone number or selecting a television channel.

By the way, the benefits of direct frequency readout are not restricted to new shortwave receivers. External readout devices for popular older shortwave receivers are available from several shortwave equipment suppliers.

SENSITIVITY AND SELECTIVITY

Sensitivity is the term used to describe how well a receiver can respond to faint radio signals and produce audio for you to hear. Sensitivity is provided by the radio frequency (RF) amplifier section of a receiver. *Selectivity* is how well the receiver can reject signals on frequencies other than the frequency you want to listen to. Selectivity is provided by the circuits in the intermediate frequency (IF) amplifier sections of a receiver. Many newcomers to shortwave think that sensitivity is the most important factor in choosing a shortwave receiver. But today's crowded bands make selectivity a more crucial factor in receiver performance.

Sensitivity is usually expressed as the input signal level (the signal delivered from the antenna to the receiver) necessary to give a signal plus atmospheric noise output from the receiver at some specified point above the internal noise produced within the receiver itself. The point usually specified is 10 *decibels* (dB). Decibels are used to express ratios between two power levels and are logarithmic; that is, a 3-dB increase in power is equal to doubling the power, but a 10-dB increase is equivalent to increasing the power ten times. The input signal level is measured in *microvolts* (μV); one microvolt is equal to one-millionth of a volt. Thus, a receiver's sensitivity might be specified as "0.5 μV for 10 dB *S + N/N*." This can be interpreted as "A half-microvolt signal fed to the receiver by its antenna will produce an output from the receiver in which the radio signal plus atmospheric noise is ten times stronger than the internal noise produced by the receiver itself." (To understand how strong receiver internal noise is, disconnect all antennas from a receiver and listen to the remaining background noise it produces. Internal receiver noise is produced by random motion of electrons within the receiver's components and circuits.) The smaller the number of microvolts specified, the more sensitive the receiver is.

As indicated at the beginning of this section, sensitivity is not as important as many new SWLs assume. This is mainly because one of the major advances in receiver technology in recent years has been in sensitivity; even simple, inexpensive contemporary receivers have sensitivity ratings comparable to professional-quality receivers of twenty years ago. Moreover, the trend among shortwave broadcasters has been to higher transmitter powers, meaning that less receiver sensitivity is required than before. For 90 percent of most listening requirements, almost any currently available shortwave radio will have adequate sensitivity. If extra sensitivity is needed for certain situations, such as DXing, there are accessory amplifiers which can amplify the signal from the antenna before it is fed to the receiver (these will be discussed in the next chapter). As a general rule, it can be stated that sensitivity ratings can be effectively disregarded when buying a contemporary shortwave radio, particularly one in the medium to upper price ranges.

Three other factors need to be kept in mind in considering the importance of sensitivity specifications. One is that small differences (0.5 μV or so) in sensitivity will have little practical effect on the signal you hear. Another consideration is that sensitivity also measures how well the receiver responds to atmospheric noise (which is a type of natural radio "signal") as well as desired radio transmissions. On many lower shortwave frequencies (such as below 5000 kHz), atmospheric noise can be stronger than many radio stations—increased sensitivity in such situations just means you'll hear the noise better instead of desired signals. The final point—and a very important one—is that a sensitive receiver can't compensate for a poor antenna system. A receiver can only process the signal delivered to it by the antenna. If the antenna is inadequate, then a good sensitivity rating is wasted. A receiver with only average sensitivity connected to a good antenna will consistently outperform a receiver with excellent sensitivity connected to a poor antenna, particularly in receiving DX stations. It's not worth paying extra for a more sensitive receiver if you can't—or won't—install a good antenna system for it. In fact, many experienced SWLs/DXers consider the antenna system to be more crucial in receiving rare DX than the receiver.

The increased sensitivity of contemporary receivers, along with increased transmitter powers used by many stations, has resulted in many of them being too sensitive in some circumstances. This problem is known as overloading and can be acute in and near international broadcasting bands during periods of peak activity. Symptoms of overloading include signals appearing on frequencies where they shouldn't and having the audio of a strong station superimposed on the audio of weaker stations; these "phantom signals" are often distorted as well. As the term implies, overloading takes place when a few strong signals "swamp" the RF amplifier section and cause it to function improperly.

An important measure of how well a receiver can handle strong signals is its *dynamic range*. This is defined as the range between a receiver's internal noise level and the signal level at which overloading effects begin taking place. Dynamic range is also measured in decibels; most communications receivers have a measurement of at least 70 dB, with a measurement of 100 dB or more considered excellent. This is one sensitivity specification which is important to consider.

There are several methods used by shortwave receiver manufacturers to reduce the susceptibility of their products to overloading. Signals from local broadcast band stations operating at 1600 kHz and below can cause overloading on lower shortwave frequencies. To minimize this, some manufacturers incorporate *high-pass filters*. High-pass filters allow frequencies above a certain point (the *cutoff* point) to pass without difficulty but block frequencies below that point. When used in a shortwave receiver, a high-pass filter with a cutoff point of 1600 kHz is included and is switched into the circuit between the antenna and RF amplifier when the receiver is tuned above 1600 kHz. In this manner the AM broadcast band can be tuned normally, yet shortwave frequencies are protected from overloading caused by stations operating there.

A method used in several earlier shortwave receivers and in a few contemporary receivers is *preselection*. In preselection, a tuning control for the RF amplifier stage is provided. This tuning control allows "peaking" the RF amplifier stage for maximum sensitivity on a narrow (such as 300 kHz or less) range of frequencies; the receiver is far less sensitive to signals outside the preselected range. Most receivers with a preselector require the listener to manually tune the preselector for best results; some better-quality shortwave receivers automatically adjust preselector tuning to correspond to the frequency tuned by the receiver.

The two most common methods of combatting overloading in contemporary receivers is an RF attenuator and RF gain control. An RF attenuator may reduce the sensitivity of the receiver by a fixed amount such as 10 or 20 dB, or it may let the sensitivity be reduced continuously. The reduced sensitivity means that strong signals are less likely to overload the RF amplifier section; however, it also means that *all* signals will be reduced in strength—strong as well as weak. The RF gain control allows the user to continuously vary the amplification (*gain*) of the RF amplifier stage in much the same manner that a volume control lets a listener continuously vary the audio level from the receiver. Both of these controls allow listeners to use only the minimum amount of RF amplifier stage gain necessary to hear a desired station and thus reduce the possibility of overloading.

With the shortwave bands becoming increasingly more crowded, selectivity is an important criterion in evaluating receivers. The major difference between most higher-priced shortwave receivers and less expensive models tends to be the selectivity specifications and options available on the more expensive units. The additional cost of a more selective receiver is often justified, particularly when one is tuning for DX signals in heavy QRM; a receiver with enhanced selectivity can frequently produce a readable signal in cases where a less selective receiver cannot.

In the previous chapter, the concept of a signal's bandwidth was discussed. Ideally, a receiver's bandwidth should be equal to exactly the bandwidth of the type of signal being received. That's not the way things work in real-world receivers, however. Suppose you tune a SW radio to 9500 kHz. It will respond to signals transmitted on 9500 kHz. However, it will also be capable of responding to signals on adjacent frequencies, such as 9498 and 9502 kHz. It can also respond to signals on 9495 and 9505 kHz and even to signals on 9490 and 9510 kHz, unless the selectivity is adequate. The strength of signals on adjacent frequencies also affects selectivity. A selective receiver may easily reject a weak or moderate interfering signal located 2 kHz from the desired frequency, yet be unable to reject a stronger signal located 3 kHz from the desired frequency.

Selectivity is measured in terms of how well the receiver can *attenuate* (reduce) an interfering signal located so many Hz or kHz away from a desired frequency range. This is known as the receiver's *bandpass*, sometimes also known as its *bandwidth*. The degree to which an interfering signal is attenuated is expressed in decibels, and is used to express the width of the receiver's bandpass at the points at which an interfering signal is reduced by 6 dB (reduced to approximately one-fourth its strength) and by 60 dB (reduced to approximately 0.0000001 percent of its strength).

As an example, assume that you want to receive an AM station operating on a carrier frequency of 9500 kHz transmitting a maximum audio frequency of 3 kHz. As explained in the last chapter, this means that the AM signal would occupy 6 kHz of frequency space. That means the receiver's bandpass should equal the bandwidth of the received signal, which is 6 kHz in this case. All frequencies outside of the 9497–9503 kHz range occupied by the AM signal are unwanted and should be rejected. Suppose the receiver has a rated AM bandpass of "6 kHz at -6 dB down." This means that any signal located outside the 6-kHz bandpass will be attenuated at least 6 dB. In a similar fashion, typical optimum receiver bandpasses for other modes include 250 to 500 Hz for CW, 1.8 kHz for RTTY, and 2.4 to 2.9 kHz for SSB. Many receivers, particularly communications models, have selectable bandpass settings for different emission modes. Such bandpasses are rated at 6-dB attenuation points in receiver advertising and manufacturer's literature.

As mentioned earlier, the ability of a receiver to reject interfering signals depends upon the strength of the interfering signal and on the strength of the signal you wish to receive. Continuing the previous example, suppose the AM signal on 9500 kHz is very strong. In such a situation, a bandpass of 8, 10, or even 12 kHz can produce excellent listening. But if there is an equally strong station on 9505 kHz, the wider bandpasses would produce more noticeable QRM than a 3-kHz bandpass. However, if the signal on 9505 kHz is significantly stronger than the one on 9500 kHz, even the 3-kHz bandpass may not prevent the signal on 9500 kHz from being overwhelmed with QRM from 9505 kHz.

The best measure of a receiver bandpass's ability to reject strong signals near the desired one is the *shape factor* of the bandpass. Shape factor is defined as the ratio of the bandpass as measured at the 6-dB and 60-dB attenuation points. For example, suppose that a bandpass is rated at 3 kHz at 6 dB down and 6 kHz at 60 dB down. In this case, the shape factor of the bandpass is 2:1.

The shape factor is perhaps the best indicator of a receiver's ultimate ability to reject QRM under difficult conditions. An ideal situation would be for a bandpass to have a shape factor of 1:1, but this is not possible in actual practice. A shape factor of 2:1 or less for various bandpasses is possible and indicates excellent ultimate selectivity; in fact, some professional-quality receivers feature bandpasses having shape factors of 1.5:1 or so. Unfortunately, the shape factor (or 60-dB selectivity) is often not mentioned in receiver advertising, manufacturer's literature, or receiver reviews appearing in magazines or SWL club bulletins.

Figure 3-4 shows a receiver's bandpass plotted graphically for a selectivity of 3 kHz at 6 dB down and 6 kHz at 60 dB down, giving a shape factor of 2:1. The horizontal axis plots the frequency, with 0 representing the carrier (or *center*) frequency of the desired signal, and the vertical axis represents the amount of attenuation in decibels. You'll note that attenuation is close to zero at the center frequency and increases away from the center frequency. The shape of the selectivity curve resembles a skirt, and thus a receiver's ability to reject adjacent QRM is known as *skirt selectivity*. Thus, a receiver with a good shape factor may be said to have "tight skirts," and a signal on an adjacent frequency that interferes with a desired one is sometimes described as "coming in under the skirts."

Selectivity is normally achieved by tuned circuits composed of inductors and capacitors. However, such circuits have difficulty in achieving the tight-skirt selectivity needed for many situations. A good way to greatly improve the shape factor of a receiver's bandpass is to use a *mechanical filter* or a *crystal filter* in the receiver's IF amplifier section. Both types of filters are based on the piezoelectric effect—the ability of certain materials to transform electrical energy to mechanical energy and vice versa. Both types of filters can be designed to pass a certain range of frequencies centered around a receiver's IF frequency (such as 455 kHz) while rejecting others. Some receiver manufacturers offer crystal filters as op-

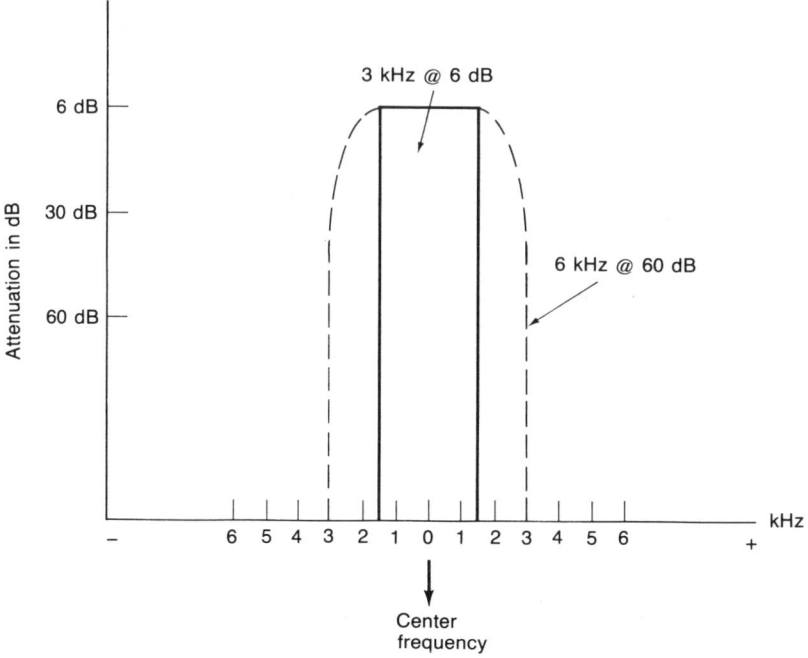

Figure 3.4 Filter bandpass with a 2:1 shape factor.

tional accessories (usually in bandwidths suitable for SSB and CW reception), while several shortwave equipment dealers will install custom crystal or mechanical filters in various receivers for improved selectivity.

Receivers having different bandpasses available will have such controls as a "wide/narrow" switch or a separate bandpass selector control. Other receivers may have different bandpasses, but these are "tied" into the mode selector switch; that is, setting the mode switch to SSB (or USB/LSB) will select the 2.7-kHz bandpass, while setting it to AM will select a 6-kHz bandpass. Having bandpass selection independent of mode is best, since it allows using the bandwidth most appropriate for a given listening situation.

A recent innovation in receiver technology is *variable bandwidth tuning* (VBT), sometimes called *passband tuning* (PBT). VBT/PBT allows the bandpass of a selectivity to be continuously varied in much the same way a volume control regulates the audio output of a receiver. For example, on a typical receiver equipped with VBT/PBT the selectivity at 6 dB down can be adjusted to any value from 2.7 kHz to only a few hundred Hz. This can be a valuable feature in the hands of an experienced SWL or DXer.

Another valuable selectivity tool is the *notch filter*. Figure 3-5 shows the selectivity curve of a notch filter. You'll notice the curve is almost the opposite of the curve in Figure 3-4, and that is how a notch filter works, when tuned to a

Sensitivity and Selectivity

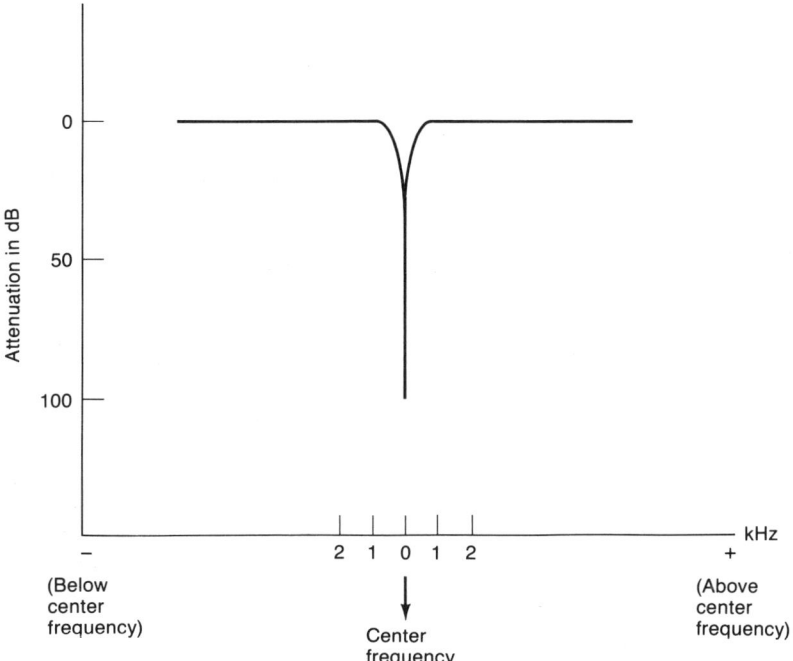

Figure 3.5 A notch filter in action.

frequency, the notch filter attenuates the signal on that frequency. The bandwidth at which attenuation occurs is very narrow (usually only a few hundred Hz). The purpose of a notch filter is to remove a "slice" of the receiver's bandpass where an interfering signal lies. For example, a 2.7-kHz bandpass may be used for receiving a SSB signal. Within the 2.7-kHz bandpass may be a CW signal. Careful adjustment of a notch filter will permit the interfering CW signal to be minimized or even eliminated altogether. Using the notch filter will somewhat degrade the audio of the received SSB signal, but it can mean the difference between being able to understand the signal or having it rendered useless by the CW QRM.

Most notch filters operate only when the receiver is in the CW or SSB mode. However, some receivers permit their notch filters to also operate in the AM mode. A notch filter can be exceptionally valuable during AM reception. When two AM signals are located close in frequency, the two signals can "beat" against each other and produce a *heterodyne*. A "het" is a piercing, high-pitched whistle that can render either or both signals unlistenable. A notch filter in such cases can eliminate the heterodyne and allow reception of the desired signal.

A "true" notch filter operates in the receiver's IF section, and is sometimes described as an "IF notch filter." A few receivers have what is known as an *audio notch filter*. This is really an advanced tone control, and blocks the audio frequency of the heterodyne whistle. An audio notch filter is not as effective as an IF notch filter.

Some receivers incorporate *audio filters* to aid in selectivity, and numerous models of audio filters are available as accessories. Like the audio notch filters, audio filters are simply elaborate tone controls capable of passing a desired range of audio frequencies, blocking an undesired range of audio frequencies (as with an audio notch filter), or tailoring a receiver's audio output to one's own preference. Audio filters are most effective for CW reception, where an audio frequency range of less than 100 Hz is needed. Audio filters are less effective for voice or music reception, and are no substitute for effective selectivity in the receiver's IF section. However, they can be useful in some situations.

With the shortwave bands becoming increasingly crowded, the selectivity capabilities and options of a receiver are important regardless of whether your listening interests are more toward SWLing or DXing. This is one area where you get very much what you pay for; the more expensive receivers generally have superior selectivity and options. An ability to choose different selectivity bandpasses independent of mode is a valuable feature worth looking for.

NOISE LIMITERS AND BLANKERS

One frequently misunderstood receiver control is the *noise limiter* or *noise blanker*. It may also be known as an *automatic noise limiter* (ANL). Many SWLs are disappointed in the performance of such controls; they are effective against certain types of noise but ineffective against other types. However, a noise limiter or blanker can be valuable in many circumstances.

Noise limiters and blankers are most effective against pulse-type noise such as automobile ignitions, switches, lightning bursts, and similar noises which can be heard for short durations. They are not effective against atmospheric static and other relatively constant noise sources. A noise limiter is a simple circuit which "shaves off" the peaks of noise pulses and reduces them to more tolerable levels. A noise blanker is a more complex circuit which actually silences the receiver during the period of the noise pulse. Both noise limiters and blankers adversely affect the audio quality of the received signal, with noise limiters generally worse in this respect than noise blankers.

A noise limiter often can only be switched on or off. The action of a noise blanker can usually be adjusted. Common adjustments include the degree to which the noise blanker operates, how fast it reacts to noise pulses, and whether it operates on wide bandwidth or narrow bandwidth noise pulses. Another use for noise blankers is reducing interference from an over-the-horizon radar system operated by the Soviet Union; this system produces a powerful pattern of signal "spikes" similar to ignition noise. Modern noise blankers can help reduce this source of QRM.

AUTOMATIC VOLUME AND GAIN CONTROLS

Many receivers today employ some form of *automatic volume control* (AVC) or *automatic gain control* (AGC). These circuits try to maintain the audio output from the receiver constant regardless of the changes in strength of the received signal. Such circuits are based around an AVC or AGC detector section which samples the strength of the received signal. If the strength drops, the AVC/AGC increases the gain of the receiver's RF and IF amplifier sections. If the signal strength increases, AVC/AGC reduces the RF and IF section gain.

Many AVC/AGC circuits have *selectable speeds*. The "speed" of an AVC/AGC circuit is how quickly it reacts to changes in the strength of the received signal. A fast AVC/AGC speed is best for SSB and CW. This is because those modes have no carrier, and no signal is present between Morse code characters or words in SSB speech; the signal levels of both modes thus change rapidly and a rapid AVC/AGC action is required. A slower speed works best for AM, since a carrier is always present and changes in signal level (due to fading, etc.) tend to take place over a longer length of time. Using a fast AVC/AGC speed on some fading AM signals can result in distorted audio.

Almost all AVC/AGC circuits can also be switched off. This is because a rapidly fading signal can be distorted by AVC/AGC action, no matter which speed is selected. In addition, maximum receiver sensitivity is obtained with AVC/AGC off.

S-METERS AND SIGNAL INDICATORS

Almost every shortwave radio comes with an *S-meter* or other signal indicator prominently displayed on its front panel. Usually such S-meters are calibrated in "S-units" from 1 to 9 with decibels indicated above "S-9" in increments such as 20, 40, or 60. Some receivers have LED signal meters, with a stronger signal level lighting up more LEDs than weaker signals.

Almost no two different shortwave receivers—even two examples of the same model—will give the same S-meter reading for an identical signal. The S-meter or other signal indicator is thus a *relative* indicator of received signal strength, not an absolute measurement such as obtained from a voltmeter or thermometer. You'll sometimes see SWLs report in various club bulletins that a signal was "20 (dB) over S-9" or that it "pinned the meter." While these are colorful descriptions of signal strength, they don't mean that a different receiver used by the same listener would indicate a similar signal strength. Listeners who own two or more receivers can verify this for themselves by tuning the same station, using the same antenna, and comparing the various S-meter readings.

This doesn't mean that an S-meter or other signal strength indicator is worthless. They can be very useful when one is using antenna tuners, preamplifiers, and similar devices to "peak" receiving equipment for maximum performance. A visual indication of how rapidly a signal is fading can suggest the proper AVC/AGC speed and even give clues as to the possible location of a station. (Signals traveling over the North Pole and its auroral region tend to have a rapid, rhythmic fading known as *flutter*.) But as an absolute, scientific method of determining a signal's strength, an S-meter shouldn't be taken too seriously.

MEMORIES AND MULTIPLE TUNING CIRCUITS

One outgrowth of digital electronics has been several receivers incorporating *memories* for various frequencies. Many of these receivers also have several different ways of tuning frequencies in memory or of tuning frequencies outside of those stored in the various memories.

You are likely familiar with the concept of a "memory" as used in other consumer electronics devices such as telephones, timers, and even microwave ovens. In SW receivers, a desired frequency is tuned and then entered into a memory (usually by pressing a few buttons in sequence). Memories are generally numbered, and a desired frequency in receiver memory can be tuned simply by turning the memory selector to the number representing the desired frequency. Often memories can store additional information, such as the mode used on a particular frequency.

Some receivers come with a *scan* function for memory frequencies or for the normal tuning circuitry. This allows the receiver to continuously tune through the memories until the scanning is stopped by the operator or a signal of a certain strength. The time spent monitoring a frequency in memory before scanning to the next varies from receiver to receiver, but is usually quite short. Some receivers allow the scanning function to also operate with the main tuning circuitry; lower and upper limit frequencies are specified, and the receiver tunes through the range. The scan rate in such cases is usually equal to the receiver's tuning speed (1 kHz, 0.1 kHz, and so forth).

Some receivers allow switching back and forth between the main tuning circuitry and frequencies stored in memory. Other receivers have what are, in effect, two main tuning circuits, which can be indicated by such phrases as "dual VFOs" or "VFO A/B" in receiver specifications or front panel labeling. These arrangements permit quickly jumping between two frequencies for such purposes as comparing signal strength or keeping track of the time on WWV/WWVH. The advantage of two main tuning circuits is that frequency, mode, and the like can usually be changed more easily than when frequencies are stored in memory. These various memory and tuning features become handy for both SWLing and DXing, regardless of whether you want to store the frequencies of your favorite stations or of rare DX "targets."

BEAT-FREQUENCY OSCILLATORS AND SSB RECEPTION

As mentioned in Chapter 2, reception of CW and SSB signals requires a beat-frequency oscillator (BFO) circuit. The output of the BFO is fed to the receiver's detector stage. Many communications receivers employ a circuit known as a *product detector*, a combination of a BFO and a special detector circuit, for enhanced reception of SSB and CW.

Simple SW receivers often have a continuously tunable BFO. To receive CW or SSB on such receivers, the BFO tuning is adjusted for the most pleasing CW sound or until SSB is resolved into intelligible speech. More advanced receivers employ fixed-frequency BFOs. Such receivers can be identified by mode selector switches or buttons labeled "CW," "USB," "LSB," "AM," and sometimes "FM" and "RTTY" (although the BFO is not normally used for AM or FM reception). On such receivers, the desired mode is selected and no further BFO tuning is required; however, sometimes the receiver tuning needs to be slightly adjusted for best audio. Some receivers have a RIT control to allow "touching up" the audio.

Even if your primary interest is in shortwave broadcasting, having fixed-mode selection instead of a tunable BFO will be useful. This is because some AM signals that are weak or suffering heavy interference are best received when tuned as if they were SSB signals. This technique will be discussed next.

EXALTED CARRIER SSB RECEPTION

Reception of AM signals on shortwave can be plagued with many problems. One involves the bandwidth required for an AM signal. If both sidebands and the carrier cannot be received, the signal may be unintelligible. Moreover, it is possible (due to the way shortwave signals are propagated) for one sideband to be received before the other one. Even though the difference in the time between reception of the two sidebands can be measured only in picoseconds or less, it is enough to "confuse" the receiver's detector stage and produce distortion. Finally, the carrier of the received signal may not be strong enough for the detector to operate properly.

However, it is possible on communications receivers to receive AM signals as if they were SSB signals. This technique is known as *exalted-carrier SSB* (ECSSB) reception. It is so termed because the signal from the receiver's BFO is "exalted" over the carrier of the AM signal. In ECSSB, an AM signal is tuned in the normal manner. The receiver's BFO is activated, either by switching it on or by choosing USB or LSB on the mode selector, and tuned so that its frequency matches (or is *zero beat* with) the frequency of the AM signal's carrier. By switching to a narrower bandpass, such as that normally used for SSB, the AM signal can be received as if it were SSB. Either the upper or lower sideband of the AM signal can be tuned in this manner. Suppose you want to hear a station on 9500 kHz, but it is suffering heavy QRM from a station on 9505 kHz. Using ECSSB, you could select a narrow bandpass (such as 2.7 kHz) and tune for the lower sideband of the station on 9500 kHz. The lower sideband would be located *below* 9500 kHz (such as from 9497–9500 kHz) and would be less affected by the station on 9505 kHz. In many cases, a station unreadable in the conventional AM mode can be made intelligible by using ECSSB.

The term "zero beat" means that the AM signal's carrier and the receiver BFO signal do not beat against each other and produce a heterodyne. To tune the BFO signal to the received signal's carrier, select "USB" or "LSB" on the receiver's mode switch or turn on the receiver's BFO (sometimes labeled "SSB/CW" or similar). You'll hear the piercing whistle of a heterodyne. Carefully tune and you'll hear the audio frequency of the whistle drop until it disappears. At that point, the BFO and carrier are "locked" together.

You may find it necessary to readjust the BFO or receiver tuning every few minutes during ECSSB reception; this is because the BFO and carrier frequency must be within a few Hz of each other, but the BFO frequency in most receivers

Figure 3.6 Sony ICF-2010 receiver uses an LCD display to indicate frequencies to the nearest 0.1 kHz.

tends to "drift" over time. More expensive communications receivers have more stable BFOs.

Exalted-carrier reception can also be used with wider bandpasses and stronger signals. The replacement carrier generated by the BFO is stronger than the one in any received AM signal. This reduces the effects of fading on the signal and helps the receiver's detector stage do its job better. One setting your author has found effective when tuning international broadcast bands is to use the normal AM bandpass but to place the mode selector to USB or LSB (usually USB) and tune for zero beat.

One recent advance in receiver technology has been *synchronous detection* which can be thought of as automatic ECSSB reception. This technique involves receiver circuitry which produces an internal carrier for exalting over an AM signals carrier and then uses either or both sidebands of the AM signal for detection. While the result is the same as the ECSSB tuning technique previously described, such a receiver can be tuned in the same manner as an ordinary AM receiver without the requirement for zero beating or selecting a sideband to tune for. Some sets employing this system have LEDs or other indicators to show which sideband is being used for detection. At this time, relatively few receivers are available using this technology. However, it has great promise for improving SW reception and seems certain to be included in future radios.

LEARNING TO USE A SHORTWAVE RECEIVER

Some new SWLs buy the most expensive SW receiver they can afford but soon become disenchanted when the receiver fails to perform up to their expectations. They wind up selling the receiver to another SWL. A few months later, the first listener notices in various SWL club bulletins that the listener who bought the receiver is using it to hear all sorts of rare DX. This isn't because SW receivers improve with age like fine wine. It's because *skill* is required to get the most out of a receiver. Such skill involves recognizing reception conditions and being able to adjust the receiver for best reception under those conditions.

One important aspect to using a receiver is to recognize that a receiver cannot do everything by itself. As mentioned before, a good receiver must be connected to a good antenna system for maximum performance. Too many new SWLs will spend hundreds of dollars on a radio but refuse to invest another dozen or so dollars on an adequate antenna. Other new SWLs will simply tune their receivers to the frequency of a station they want to hear and then expect the station to come in loud and clear; when it doesn't, they assume that the receiver is defective or that they were misled by the dealer or other SWLs. In some cases, they try to hear the station when it's not even on the air! In other instances, they try to hear the station at times when reception conditions will not permit it to be heard. For example, no matter how good a receiver or antenna one might use, it will be impossible to hear a station from Brazil on 90-meters at noon from a listening site in central North America.

Besides a knowledge of what is possible to hear, using a shortwave receiver well takes practice. It takes time to understand the effect of various controls upon what you hear. For example, using AVC/AGC and a noise blanker properly can make the difference between an understandable signal or unintelligible noise. Another important control to use properly is the RF gain control. For example, many SWLs find SSB reception best when they turn the receiver's volume control (sometimes labeled *AF Gain*) to maximum and use the RF gain control as a "volume" control. This technique does not work for all receivers, but is the sort of technique (like tuning for zero beat during ECSSB reception) that a SWL must learn from experience and practice.

Experience will also teach you how to best use the bandpass selections supplied with your receiver. It is possible, for example, to tune a 6-kHz-wide AM signal by using a 2.7 kHz bandpass filter without resorting to ECSSB. This is done by receiving the original AM signal carrier and a *portion* of one of the sidebands. This results in "clipped" audio and a loss of fidelity, but it can mean the difference between understanding a signal and not being able to identify it at all. Similarly, SSB signals can be received by using a 1.8-kHz bandpass intended for RTTY reception. The audio fidelity is reduced, but it can let you at least understand signals otherwise lost in QRM. Learning how to use various bandpasses takes time, and can't be accomplished overnight.

An important part of any receiver is the SWL's *ears*. It takes experience to recognize a weak signal buried in QRM that might be improved through use of a narrower bandpass or ECSSB. DX signals, by definition, are not powerful, easily noticed signals. Less experienced SWLs may tune past weak, QRM-laden DX signals that catch the attention of a more experienced listener. Once the more experienced SWL notices such a signal, he or she then can make best use of the receiver to bring the signal up to an audible, understandable level.

Using an SW receiver, especially a communications model, is like driving a sports car or using a telescope. A new set of skills is necessary to get the most out of it, and those skills are acquired only with time and effort.

WHICH RECEIVER TO CHOOSE?

There is no single receiver best for everyone. The receiver that is right for you depends upon your interests and level of experience.

If you're starting out in SWLing, a relatively simple portable SW radio may be a good bet as long as it has direct frequency readout and is able to receive SSB/CW. Such a set will allow you to sample the various activity you can hear on shortwave, to gain experience, and to define your listening interests and needs more precisely. If your main interest turns out to be listening to major international broadcasters, such a portable will likely be all the receiver you ever need.

If, on the other hand, SWLing fails to hold your interest, you haven't made a major investment and you still have a radio useful for traveling or emergency use.

One of the best sources of information about receivers is DX club bulletins. These bulletins indicate which receivers were used by members for the receptions reported in them. If several members are using the same model receiver, this normally indicates that the receiver is a good performer or offers value for the money. Club bulletins also run reviews of various receivers and comments by members on receivers. Such reviews and comments, while not always professional, tend to be less biased than those appearing in magazines and similar publications which depend on advertising dollars from receiver manufacturers. Notable exceptions are the reviews prepared by Lawrence Magne that appear in *The World Radio TV Handbook* and *Radio Database International*, which are currently issued annually. Addresses and sources for these publications can be found in the appendix.

While the emphasis on this chapter has been on selecting a new receiver, used models of popular shortwave receivers can be located through SWL club bulletins, shortwave equipment dealers, and amateur radio magazines and publications. Some outstanding bargains can be found with a little patience.

If the choices facing you when selecting a receiver seem daunting, you can at least take comfort in the fact that your dollars buy more and better performance than ever before. While making the *best* selection can be a problem, it is difficult to make a *bad* choice today.

4

ANTENNAS AND ACCESSORIES

Antennas may be the most misunderstood topic among SWLs. A quick glance through various SW club bulletins and commercial publications will show there is no shortage of articles and other information (and sometimes misinformation) about antenna systems. Yet many SWLs still use an antenna system that is inadequate or ill-suited for their needs. It's not uncommon to find some SWLs who spend hundreds or even thousands of dollars on a receiver and then use for an antenna any spare wire they can find around the house. In such cases, the SW receiver can't deliver the performance listeners expect; by saving a few dollars on the antenna, they waste the receiver capabilities they paid so dearly for. In other situations, listeners sometimes *overpay* for antenna systems and devices that can't perform as well as a simpler antenna costing much less.

A similar situation exists with receiver accessories. Properly selected accessories can greatly enhance the "DX catching" ability of less expensive receivers and make an outstanding receiver even better. However, poorly chosen or incorrectly used accessories can easily *decrease* the performance of even the best SW receiver.

With both antennas and accessories, it's often true that the most expensive or complex solutions aren't the best. A little knowledge about the two will let you decide on simple, cost-effective enhancements for your SW receiver.

SOME SIMPLE ANTENNA THEORY

The subject of antenna theory can be complex—just look at any book on antennas published for radio amateurs or communications engineers. That is because such users of antennas are usually concerned with transmitting with a given antenna as

Some Simple Antenna Theory

well as receiving. An improper antenna in such situations can easily cause damage to a station transmitter. Fortunately, antenna requirements for receiving are much less demanding, and the only penalty for using a poor antenna is poor reception. (Of course, any receiving antenna must be installed and used with safety considerations in mind. These will be discussed later.) It's possible to install and use a SW receiving antenna in situations where a SW transmitting antenna would be difficult or impossible to install or use. Thus, living in an apartment, condominium, dormitory, or even a high-rise building is no bar to enjoying SWLing with an efficent antenna. As long as you prevent stray, unwanted electrical charges (lightning, contact with power lines, and the like) from entering your receiver via the antenna, you may connect whatever type of antenna you wish to your receiver without worrying about damaging it.

The basic principle behind an antenna is simple. Radio waves traveling through space strike an antenna and cause very feeble electric currents to be produced (or *induced*) in the antenna. These faint currents are then fed from the antenna to the receiver. This simple description of how an antenna works intuitively suggests that a larger or longer antenna works better than a smaller antenna, since the larger or longer antenna would logically seem to be able to "capture" more of the energy present in radio waves. It also suggests that an antenna should be clear of metallic obstructions which could absorb radio wave energy before the radio waves strike the antenna. The second observation is always correct; every antenna should be "in the clear" as much as possible. The first observation is sometimes correct, but not always. Depending on the frequency range being tuned, a shorter, smaller antenna may outperform a longer and larger one.

The reason why bigger is not always better with shortwave antennas is that an antenna gives best performance when it is *resonant* at the frequency being tuned by the receiver. Whether or not an antenna is resonant depends on the *wavelength* of the radio signal being tuned. All radio signals travel as a series of waves having positive and negative peaks. Figure 4-1 shows a typical radio wave; the waveform of a radio wave is the type known as a *sine* wave. The wavelength

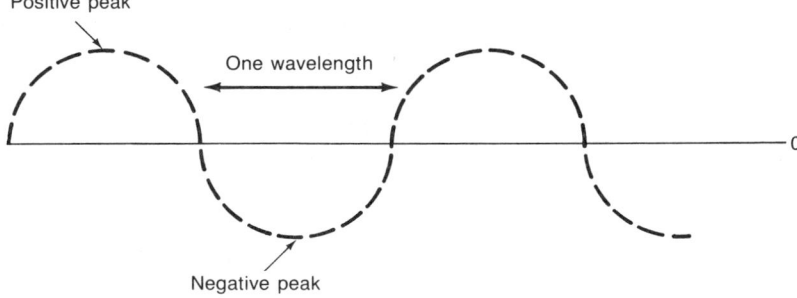

Figure 4.1 A radio wave and wavelength.

of a radio wave is defined as the distance between the positive peaks of the wave, as shown in Figure 4-1. As mentioned in Chapter 2, wavelength is measured in meters. Thus, the wavelength of radio waves in the 3500–4000 kHz range is around 80 meters. As signals increase in frequency, their wavelength decreases.

An antenna is resonant at a given length when its length is such that it lets an electric charge travel from one end of the antenna to the other and then back again in the time necessary for the radio waves to go through a complete cycle (from one positive peak to the next positive peak, as shown in Figure 4-1). Since the electric charge travels the length of a resonant antenna twice, you might suspect this means that a resonant antenna should be half a wavelength long at the frequency you wish to receive. This is very close to the truth; a resonant antenna for the 80-meter amateur band should be approximately 40 meters in length.

The definition of resonant antenna depends upon the *time* it takes for a charge to travel to the other end of an antenna and back. This is defined as the *electrical* length of the antenna, and can be much different from the physical length of the antenna. The physical length of the antenna is only one factor affecting the electrical length of an antenna. It is possible to use coils of wire (known as *inductors*) and tuning capacitors to alter the length of time it takes for a charge to travel the length of the antenna and back. This means that a resonant antenna can be longer or shorter than a half-wavelength, and that a single antenna length (such as 100 feet) could be "tuned" to resonance over a wide range of frequencies by using a tuning device constructed from variable inductors and capacitors. In fact, this type of antenna system—a random-length wire fed to the receiver through an antenna tuning device—is an excellent choice for SW reception and may well be the most popular SW antenna system in the world.

A resonant antenna is important for transmitting, since a nonresonant antenna would reflect power back to the transmitter and cause damage to the transmitter. It's important in receiving antennas because maximum signal is transferred from the antenna to the receiver at resonance. This involves *impedance*. Impedance can be thought of as a circuit or device's opposition to the flow of a current. When two electrical circuits or devices are connected, the maximum power transfer takes place when the impedances of the connection points are equal. In our case, this means that the impedance of an antenna should equal the impedance found at the receiver's antenna input connections. Impedance is measured in *ohms*, which are represented by the symbol Ω.

A resonant antenna for a given frequency that is one-half wavelength long has an impedance in the range of approximately 35 to 70 Ω. For this reason, almost all SW receivers have an antenna input impedance of 50 Ω. This is an impedance value near the middle of typical resonant antenna impedances, and any differences in values between the receiver and antenna aren't significant in actual practice. Thus, maximum power is transferred between the receiver and antenna, and the receiver has the most signal available to work with.

Untuned, nonresonant antennas have impedance values which can vary widely but are usually in the neighborhood of several hundred Ω or even higher.

When such an antenna is connected to a receiver's 50-Ω antenna input, the mismatch is severe, and much of the signal developed in the antenna cannot be transferred to the receiver and is "lost." To partially compensate for this, some receivers have a *high impedance* antenna input (sometimes labeled *high-Ω* or *high-Z*) in addition to the 50-Ω antenna input. These high-impedance inputs have impedances of several hundred Ω and are better able to match untuned antennas. However, there usually remains an impedance mismatch and loss of signal. This can be verified by using the same random-length antenna and connecting it without a tuner to the receiver's high-impedance input and then switching it through a tuner to the 50-Ω input. Tuning the same stations and comparing the signals from the two antennas will show that a significant improvement in signal strength can be obtained by using the tuner and 50-Ω input. While the difference isn't important in most situations, it can mean the difference between hearing a weak DX station or just hearing noise.

The 50-Ω connector on receivers is designed to accept a coaxial cable plug, while high-impedance connectors are usually of the "spring" or screw terminal type designed for direct connection to a random wire. There is usually no problem in using tuners, since most have an input connection for random wire (such as screw terminals) and an output for coaxial cable. A few tuners, however, use "phono" plug connectors (often found on stereo equipment) as a cost-saving measure. Adapter plugs for such situations are available from electronics parts and shortwave equipment dealers.

A SIMPLE 1.6–30-MHz OUTDOOR ANTENNA

Figure 4-2 shows an antenna system useful for reception from 1600 to 30000 kHz. You can find all the necessary parts, with the possible exception of the antenna tuner, at a local electronics store. All components, including the tuner,

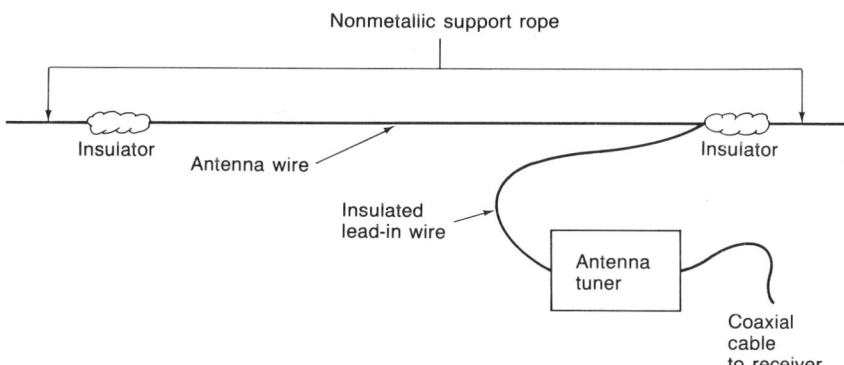

Figure 4.2 Diagram of a useful all-band antenna system.

can be obtained from shortwave and amateur radio equipment. You'll need two antenna insulators (either the "egg" or "dog bone" type), nonmetallic support rope, insulated wire to connect the antenna to the receiver (called "lead-in"), and bare copper wire. The best bare copper wire is the stranded variety in #12 or #14 gauge; solid wire tends to "kink" more than stranded. Insulated wire can be substituted for the bare copper wire, since radio waves penetrate plastic insulation easily and lose no strength.

Special attention must be paid to safety before attempting to install this or any other antenna. Carefully survey the area in which you plan to install the antenna (including the path the lead-in wire will take). *Make certain that there is no possibility of any contact between any part of the antenna system and power lines, lighting fixtures, telephone lines, air conditioning units, transformers, or other voltage sources.* Each year in the United States, several people are electrocuted while attempting to install various types of antennas. Almost each such death could have been prevented by advance planning. Fortunately, almost no deaths result from attempts to install SWL antennas. Let's keep it that way!

Antenna insulators come with two holes in them. Pass the nonmetallic support rope through one hole of each insulator and secure the support rope tightly to each insulator. The support ropes can be attached to various supports such as trees, poles, sides of buildings, and the like. The other hole on each insulator is where the wire making up the antenna itself is connected. Approximately two feet of the antenna wire should be passed through this hole and looped through it a few times. The remaining wire is then twisted around the antenna wire or insulator. The length of the antenna wire itself is not critical, but a length between 50 and 100 feet is a good choice. You won't notice a major difference in reception using longer or shorter wire lengths unless the length is significantly shorter (25 feet or less) or longer (over approximately 250 feet or more).

The crucial connection involves attaching the insulated lead-in wire to the antenna wire. The attachment can be made wherever it's most convenient for you—at either end of the antenna or even in the middle. Note that plain insulated wire is used instead of shielded cable (such as coaxial cable), and the lead-in itself absorbs additional radio energy from signals. The lead-in wire's connection to the antenna wire should be *soldered*, not merely wrapped or mechanically connected (as with an "alligator"-type spring clip). This is because a mechanical connection will quickly become loose or deteriorate, causing a high-resistance connection (or even a broken connection) between the lead-in and antenna wires. The result is an antenna system that performs much less effectively than it might.

The necessary solder connection is performed before you install the antenna. A good solder connection consists of a secure mechanical connection which is then soldered properly. To make the mechanical connection, select the area on the bare antenna wire you wish to connect the lead-in to. Using fine sandpaper, gently rub a two-inch-long area at that point until the copper is bright and shines like a new copper cent. Remove any stray particles and grit from the area, using a damp cloth. Remove about four inches of insulation from the lead-in

A Simple 1.6-30-MHz Outdoor Antenna

wire and repeat the process on the exposed wire. Remember to use a *fine* sandpaper and rub *gently* since the wire can be damaged if you're too rough.

Wrap the lead-in wire tightly around the antenna wire. A good method is to leave approximately one-half inch of stripped lead-in wire "below" the point where the lead-in first makes contact with the antenna wire and wrap any remaining exposed lead-in wire around this section. Some SWLs prefer to wind by looping three or four turns of exposed lead-in around the antenna wire and then looping the exposed lead-in around the half-inch section. They then resume looping the exposed lead-in around the antenna wire and repeat the process as necessary. A little practice will soon show which method works best for you. The only requirement is that the resulting mechanical connection be both tight and secure.

The connection *must* be soldered by using only rosin core solder; the acid core solder used for metal work will cause electrical connections to corrode. A soldering iron of 30-40 watts is usually adequate, although a higher-wattage iron or gun may be necessary if you're making the connection outdoors on a cool day. How to make a good solder connection is outside the scope of this book, but the secret is to apply heat to the connection to be soldered rather than the solder itself. Don't melt solder with the iron and then let it drip or flow over the connection. Instead, apply the iron directly to the lead-in/antenna wires connection; it's best to heat the "bottom" of the connection. Let the connection heat for several seconds and then apply solder to the "top" of the connection. Keep applying solder until the entire connection is coated with solder and then remove the iron. Don't touch the connection for about a minute or so as it cools—this is necessary to let the solder "set" properly. If you've done a good soldering job, the result will be a connection covered by bright, shining solder. Dull, pitted solder indicates a poor connection that won't conduct properly and will likely come loose. (If you've never made a soldering connection before, practice with some scrap wire before attempting the connection of the lead-in to the antenna.) After the connection has cooled, wrap it in weather-resistant tape such as electrical tape.

Unlike some types of antennas, this system will show few (if any) directional characteristics. Thus, the direction in which you install it will have negligible impact on what you can hear. The prime consideration in deciding the location of the antenna must be safety, and once this is satisfied, the only real criterion is your convenience in installation. Using the nonmetallic support ropes, hang the antenna between any two supports (but *never* power or telephone poles). Typical supports include a house and adjoining garage, utility building, or tree, or the antenna can be suspended between two trees or nonmetallic poles (such as surplus wood telephone poles).

The lead-in is brought into the house to be connected to the antenna tuner. Most lead-in wire can be easily slipped under a closed window without difficulty. Some packaged antenna kits for SWLs include what is known as a "window feedthrough" or similar term. This is a short (usually less than a foot) length of flat wire, similiar to "twinlead" cable often used with TV antennas, that has metal connectors at each end. The idea behind this is to slip the window feed-through under a window sill and close the window. The lead-in is then connected to the metal connector outside the window, and the antenna tuner is then connected to the connector inside the window. Generally, a window feed-through strip is more trouble than it's worth. For best results, you'll have to solder both connections to it and you'll have two outside connections to corrode or come loose instead of one. Moreover, it's difficult to imagine a situation in which the window feed-through could fit under a closed window but a simple lead-in wire could not. Given the the dubious value of this item, it's a mystery why some manufacturers continue to include it in their antenna packages.

Almost all antenna tuners are designed for amateur radio purposes. However, these all work well for SWLs. The best type to select is what is usually described as a "random wire" tuner. Happily, this type is simple and inexpensive. A random wire tuner usually has only two controls, labeled *inductance* and *capacitance*. The "inductance" control moves in steps (like a channel selector on a television), while the "capacitance" control is continuously variable (like a volume control). On the rear of the unit is a pair of coaxial input connectors (described as "SO-239" connectors in manufacturer's literature). The antenna lead-in is connected to one SO-239 and the other is used to connect the tuner to a receiver using a short length of coaxial cable with input (PL-259) connectors. Connecting the lead-in to the tuner is simple. Remove about three or four inches of insulation from the end of the lead-in and "fold" the wire together to make a "plug." This plug is then inserted into the antenna input SO-239.

Using the tuner is simple. The "inductance" control is set until the station being tuned is loudest or the background noise is greatest. The "capacitance" control is rotated for peak signal strength or background noise. As an analogy, "inductance" can be thought of as a "coarse tuning" control, while "capacitance" is like a "fine tuning" control. With practice, you'll soon know which setting of the "inductance" control is best for a given frequency range and you won't have to hunt for it.

This antenna is very easily to construct, yet will give years of satisfactory service if carefully built and installed. If you have the space for it, this antenna is probably the best choice, all things considered, for SWLing and DXing in the 1600–30000-kHz range.

COAXIAL CABLE, GROUNDS, AND LIGHTNING ARRESTORS

The uses and limitations of coaxial cable and lightning arrestors are often poorly understood and misused by SWLs. In the case of the former, such a situation costs only money and performance. In the latter case, it could cost you your property or life.

Some readers might be wondering why coaxial cable was not used as the

Coaxial Cable, Grounds, and Lightning Arrestors

lead-in for the all-band antenna just discussed. Coaxial cable was originally developed for transmitting applications. Its construction involves a center conducting wire surrounded by foam or plastic insulation. This insulation is in turn surrounded by a metal braid, which is sometimes known as a *shield*. This metal braid is surrounded by an outer layer of plastic or rubber insulation. "Coax" comes in different sizes and types, and is rated by its ability to handle transmitter power and how much of that power is lost between the transmitter and antenna by the coax. Common types of coax include RG-58A/U, RG-8U, and RG-213. The impedances of these types is 50 Ω.

The principal advantage claimed for coax is that it prevents noise and unwanted signal pickup because it is "shielded" by the metal braid, which prevents radio signals from entering or leaving the coax. In addition, it is also claimed that coax prevents loss of precious signal strength between the antenna and receiver. Actually, these claims are true when coax is used with certain types of antennas such as the dipole (discussed later in this chapter). But when used with most other antenna types, particularly random-wires, coax's advantages are wasted.

This is because coax must be properly *terminated*, meaning it must be connected to a so-called "balanced" antenna (such as a dipole) or to *ground*. In electronics, "ground" does not necessarily mean the earth itself; it means a point of zero voltage (which the earth is) to which an electrical circuit can be connected without affecting operation of the circuit. If coax is not properly terminated, the shield has no effect; it even acts as a random-wire antenna itself. Suppose coax was substituted for insulated wire as a lead-in for the all-band antenna, and the center wire of the coax was connected to the antenna wire. If the coaxial cable is terminated in a PL-259 input connector and connected to an antenna tuner, the metal braid will serve as another random-wire antenna rather than a shield. If the center connector of the coax is connected to the antenna tuner directly without using a PL-259, the braid simply "floats" electrically and does nothing—it delivers no signal to the tuner, and it doesn't prevent unwanted signals from penetrating the coax.

Despite the preceding, one sometimes runs across SWLs who swear that the performance of their random-wire antenna improved markedly when they switched from ordinary lead-in wire to coax. In almost every case, the improved performance is the result of the SWL's finally making a proper connection between the antenna itself and the lead-in when he or she switched to coax. Many SWL equipment suppliers stock coax and recommend it to their customers. There's a good reason for this, called profit, but it has nothing to do with performance. If you're using a random-wire antenna, the only coax you need is a short length to connect the tuner and receiver. Otherwise, forget about coax unless you're using an antenna whose design requires its use.

Grounds are also a source of much confusion. For years, SWLs have been urged to "properly ground" their receivers by running an insulated wire from their receiver's ground terminal to a cold-water pipe or metal rod driven into the earth. Even contemporary SW receiver manuals recommend this practice. SWL "folk wisdom" holds that grounding a receiver increases its performance and protects the receiver from lightning. Neither is true. Grounding was often necessary back when receivers used vacuum tubes, but today's all solid-state receivers are adequately grounded through the AC power line. Trying to obtain a ground by the conventional method does not enhance receiver performance and in fact may *degrade* it by bringing in electrical noise. Far more dangerous is the myth that grounding a receiver protects it against lightning strikes entering from the antenna. *Nothing* can protect a receiver against a lightning strike entering from the antenna, and that includes a category of devices known as *lightning arrestors*.

The term "lightning arrestor" is dangerously misleading, since it implies that using one will protect a receiver—and the room in which it is located—in the event of a direct or nearby lightning strike against an antenna. This is totally false. *There is only one way to protect a receiver against lightning and that is to disconnect the antenna and unplug the AC line during thunderstorms or if it is to be left unattended during thunderstorm season.* There is *no* other way to adequately protect the receiver from lightning. *None.*

This is not to imply that all such devices are worthless. Many modern receivers can suffer damage from static electricity charges which can build up on an antenna (from wind-driven particles such as sand and the like) even during fair weather. In such cases a "lightning arrestor" can protect a receiver by draining away those charges to ground. For this protection to be effective, the arrestor must be connected to a ground rod to provide a path for the stray electric charges to ground. Some devices of this type (sometimes known as "lightning protectors") use a gas-filled cylinder "fuse" which pops when the start of a voltage surge is detected. Like a fuse, the cylinder must be replaced if it blows. But if you have invested considerable money in a receiver, this type of device can provide inexpensive protection against damage from static electricity.

Don't deceive yourself into thinking that any "lightning arrestor" or "protector" actually does what the name implies. In the event of a thunderstorm, unplug the receiver from the AC line (to prevent voltage spikes from entering the receiver via the home power lines) and disconnect the antenna from the receiver or antenna tuner. Some listeners go so far as to remove the lead-in from inside their homes and even arrange outside hooks or other supports to hang the lead-in wire for when the receiver is not in use. To disconnect the antenna and AC power line each time a thunderstorm might threaten can be a bother. But it's less of a bother than having to buy a new receiver to replace one destroyed by lightning.

THE DIPOLE ANTENNA

If you're interested in monitoring a few narrow frequency ranges about 500 kHz wide instead of the entire SW spectrum, the *dipole* is an excellent choice. The dipole is only slightly more difficult to construct than a random wire antenna such

The Dipole Antenna

as the one in Figure 4-2, yet the dipole outperforms a random wire in the frequency range it is designed for. An added advantage is that a dipole has an impedance at resonance of approximately 75 Ω, which means it can provide a good match to a receiver without using an antenna tuner. If your listening interests are restricted to a few bands (such as those used for amateur radio or international broadcasting), the dipole is an excellent choice.

The basic construction of a dipole is shown in Figure 4-3. It is a length of wire that is one-half wavelength long at a desired frequency. The wire is "split" into two sections, each one-quarter wavelength long, by an insulator inserted into the middle of the half-wavelength-long wire section. The dipole is connected to the receiver by 50-Ω coax. The center wire of the coax is connected to one of the quarter-wavelength sections, while the metal braid of the coaxial cable is connected to the other quarter-wavelength section. It doesn't matter which conductor (center wire or braid of the coax) is connected to which quarter-wavelength section of the dipole, but each must be connected to a separate section (and one section *only*, and the two sections must not be connected together in any way. The coax can be connected to the receiver's 50-Ω antenna input by using a standard PL-259 input plug. The dipole gives good performance and a suitable impedance match at the resonant frequency of the half-wavelength wire plus about 300 kHz above and below the resonant frequency. Performance will decline and the impedance match will worsen the further you tune away from the resonant frequency of the dipole.

The length of the half-wavelength section for a dipole intended for a given frequency can be found by the following formula:

length in feet = 468 ÷ resonant frequency in MHz

Suppose that a dipole to tune a 500-kHz range centered around 10 MHz is

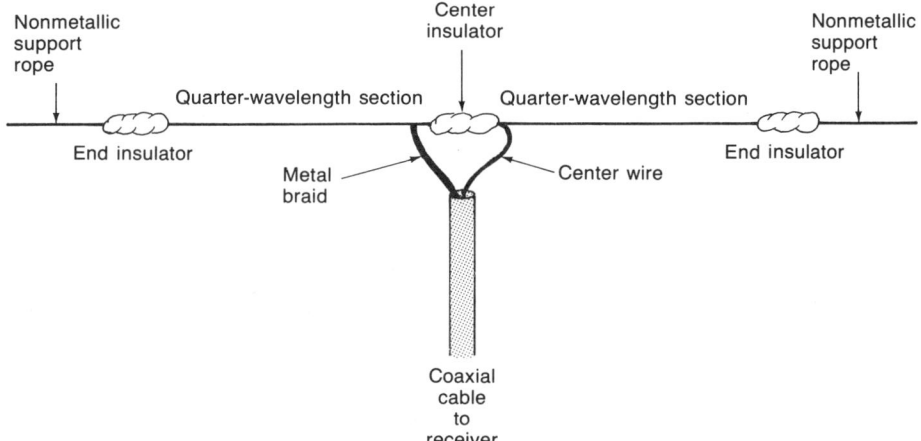

Figure 4.3 Construction details of a single-band dipole.

desired. The length in feet would be found by 468 ÷ 10, giving a length of 46.8 feet; each quarter-wavelength section would thus be 23.4 feet long. Similar calculations will give the length for dipoles designed for other resonant frequencies found from 1.6 to 30 MHz. Table 4-1 gives the length of dipoles for the major broadcasting and amateur radio bands. The resonant frequency of the dipole is indicated, and you'll notice that a frequency near the center of each band has been selected.

Construction of a dipole follows the same general mechanical principles as for a random-wire antenna. When measuring and cutting the wire for each quarter-wavelength section (or for the half-wavelength section, if you prefer to cut it in half to obtain the quarter-wavelength sections that way), allow extra wire for connections to the end and center insulators. Approximately a foot extra should be adequate. The coax braid connection is often the most difficult part of building a dipole. Start by removing approximately one foot of the outer insulation from the coax so that the metal braid underneath is revealed. Make a hole in the metal braid just above the point where the braid meets the outer insulation. Make the hole large enough so that the inner layer of insulation can pass through the hole. Carefully pull the inner insulation covering the center wire out through the hole using tweezers, a nail, or some other tool. Continue until the inner insulation has been completely pulled through the braid. Then twist the braid until it forms a "wire" suitable for connection and soldering to one of the quarter-wavelength sections. You'll notice that the center wire will probably wind up being much longer than you need; clip off any unwanted excess.

Take particular care that the braid and center wire of the coax do not come into contact at the center insulator (or anywhere else). If they do make contact with each other, the dipole becomes, in effect, a random-wire antenna. Electrical

TABLE 4-1 Dipole Lengths for Various Bands

Length (in feet)	Frequency Range
195	2300–2500 kHz
141	3200–3400 kHz
125	3500–4000 kHz
96	4750–5060 kHz
76.5	5950–6200 kHz
65.5	7000–7300 kHz
48.75	9500–9800 kHz
39.5	11700–12000 kHz
31.5	14000–14350 kHz
30.75	15100–15450 kHz
26.3	17700–17900 kHz
22	21000–21450 kHz
21.7	21450–21750 kHz
16.5	28000–29700 kHz

The Dipole Antenna

tape can be used to insulate them from each other if necessary. Connect the braid and center wire to the quarter-wavelength sections as close to the dipole's center insulator as possible. A good technique is to loop the appropriate conductor (braid or center wire) together with the quarter-wavelength section wire through the insulator hole and then solder together.

Construction of a dipole can be greatly simplified through the use of a center insulator designed specifically for dipoles. These insulators offer easy connection of each quarter-wavelength section and have a SO-239 coax input connector built in. With such a connector, there is no need to do any cutting or soldering of the coax; a PL-259 connector will directly "fit" the insulator by simply screwing it on. Center insulators designed for dipoles cost only a few dollars and are well worth the time and aggravation saved. SWL and amateur equipment dealers carry them.

A device sometimes confused with a special dipole center insulator is the *balun*. "Balun" is a contraction for "balanced to unbalanced," and is a device used to reduce unintended radiation from the coax line when a dipole is used for transmitting. Baluns are more expensive than special dipole center insulators and have no real benefit for SWLing purposes.

Unlike a random wire, a dipole shows some directional characteristics. A dipole gives best reception from directions located "broadside" to it. Figure 4-4 shows the reception pattern of a dipole. However, these effects are seldom noticeable on frequencies below 15000 kHz. For transmitting purposes, a dipole should be located at least one-half wavelength (at its resonant frequency) above the ground. This is not so critical for receiving, but best performance results when a dipole is installed as high as possible.

A dipole can be installed in different configurations, as shown in Figure 4-5. Figure 4-5(a) shows a so-called "inverted-V" dipole. The center insulator is the highest point of the antenna and the rest of the antenna slopes downward. The quarter-wavelength sections should be separated from each other by at least 30 degrees in an inverted-V. The inverted-V is a favorite of many SWLs and gives excellent DX performance. Figure 4-5(b) shows an "inverted-L" antenna, while Figure 4-5(c) shows a "sloper" dipole. The latter two designs are useful in situations where "normal" dipole support structures are not available.

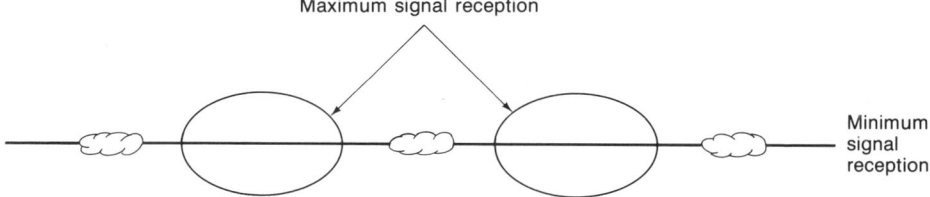

Figure 4.4 Reception pattern of a dipole.

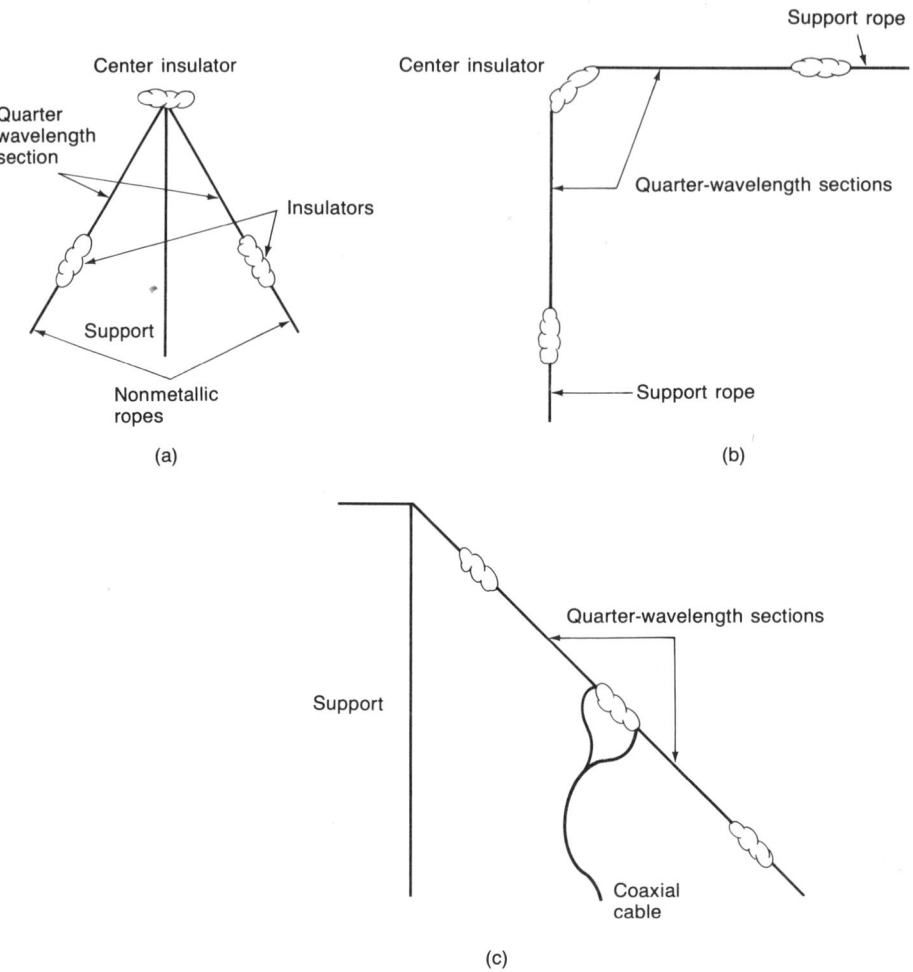

Figure 4.5 Three different methods of configuring a dipole.

If reception on several different bands is desired a *multiband dipole* can be constructed. Figure 4-6 illustrates multiband dipoles. As you can see, they are really several different dipoles sharing a common center insulator and coaxial cable. When a specific band is tuned, the "extra" dipoles have such a high impedance that they are "tuned out," and only the dipole resonant at the frequency being tuned delivers a signal to the receiver. There are two precautions to be observed in constructing a multiband dipole. The first is to make sure the various quarter-wavelength sections do not touch each other *except* at the center insulator; if they do, the antenna functions as a random-wire antenna. The other is that the quarter-wavelength sections should "match" with each other, as shown in Figure 4-6. For example, if you have a multiband dipole for 20- and 15-meters,

Other Types of Antennas 73

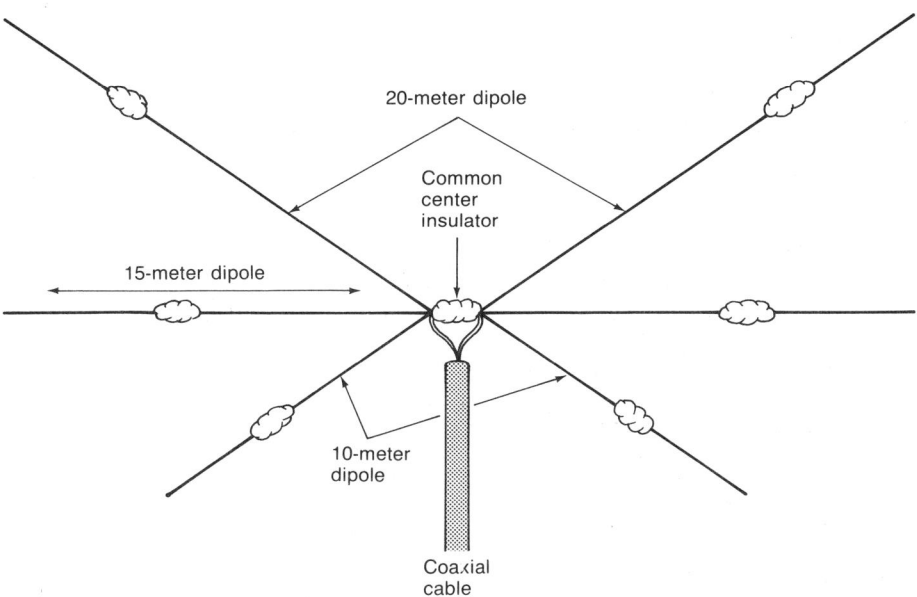

Figure 4.6 Construction of a multi-band dipole.

one of the quarter-wavelength sections for 20-meters should not be on the same line with a 15-meters section, nor should they be combined to form an angle.

There are several other multiband dipole types. Consult an antenna reference guide for radio amateurs or copies of amateur radio magazines for additional designs.

OTHER TYPES OF ANTENNAS

One approach to covering several different bands with a single antenna system is the *trap dipole*. A "trap" is an inductor (coil of wire) enclosed in a weatherproof plastic housing. A trapped dipole has a center insulator and two sections through which various inductors are spaced. These traps make the antenna resonant on various bands much like a multiband dipole. A trap dipole is also shorter than a comparable multiband dipole, since the inductors electrically "shorten" the antenna. Although it is possible to construct a trap dipole yourself, most SWLs purchase commercially available trap antennas. These systems cover the major international broadcasting bands and usually 60-meters. These systems tend to be significantly more expensive than multiband dipoles you could construct yourself and, despite sometimes inflated advertising claims, do not perform significantly better than "standard" dipoles of the single- or multiband variety. However, they are more compact than a multiband dipole and are much simpler to construct.

Another antenna type used by some SWLs is the *vertical*, which can be constructed or purchased commercially. The standard vertical is a vertical metallic rod one-quarter wavelength long mounted above a set of *radial wires* which extend outward from the point where the vertical rod is mounted. The radial wires may be buried underground if the vertical is ground-mounted or extended outward with supports if the antenna is mounted on the roof of a building. Like dipoles, verticals can be electrically shortened by including traps in the vertical rod. The "ground plane" antenna used for CB radio and scanners is an example of a vertical. Verticals have been favored by many amateur radio operators interested in DX, since a vertical responds well to radio signals arriving at a low angle above the horizon; this happens to be the angle from which many DX radio signals arrive. Another advantage of verticals is that they can often fit into less space than an equivalent dipole.

Verticals have some shortcomings, however. For best results, a vertical requires an adequate radial system; too few radials can result in poor performance. Moreover, many verticals require guy wires to prevent the antenna from being blown over by winds.

The only commercially available verticals for SWL purposes cover just the major international broadcasting bands and are quite expensive for the performance they deliver.

ANTENNAS FOR LIMITED SPACE AND DIFFICULT SITUATIONS

Many SWLs today find themselves living in condominiums, apartments, or residential areas in which the erection of outside antennas is difficult or prohibited. Such restrictions have effectively put many amateur radio operators off the air or restricted them to operation on the VHF/UHF amateur bands. The situations for SWLs is not so bad since, as mentioned at the beginning of this chapter, antenna requirements for receiving are not as critical as those for transmitting.

Some SWLs attempt to get by with an indoor antenna. This can work *if* the building is a wood or simple masonry structure. Radio waves can penetrate wood well, and a single brick wall will not result in significant loss of signal. The situation changes drastically in a building constructed with steel and reinforced concrete, however, as such buildings absorb a great deal of radio energy, and an antenna located inside such a structure would be virtually useless. The same situation also applies to antennas located inside mobile homes. Homes and other structures with aluminum siding may or may not present similar problems; it depends on whether there is another path for radio energy to enter the structure (such as through a nonmetallic roof).

The traditional installation spot for an indoor antenna is the attic. You may even have room to install a dipole, especially for the higher-frequency bands. But more typically an attic antenna is a random wire connected to an antenna tuner. Running the lead-in wire from the attic to the room where the receiver is located can be unsightly and a problem. (Kids love to tug on it, and adults trip over it.)

One inelegant—but often effective—solution is to simply leave the random wire scattered out of sight somewhere in the room where the receiver is located. Possible hiding places include a closet, under furniture (such as a bed or sofa), or even behind the desk, table, or other piece of furniture on which the receiver is placed. Don't laugh—such a "mess," when used with an antenna tuner, can give very satisfactory reception of major international broadcasters and some DX.

Some listeners have had good results with so-called "invisible" antennas. Such antennas are not really invisible, but are designed so that they are very difficult to notice. A very small wire gauge is used, with #28 and #34 being favorites (these are sometimes known as "magnet" wire). Insulators are hand-constructed from clear plastic tubing or sections. If coax is required, special small-diameter coax such as the RG-174/U type is used. The entire antenna structure is light enough to be supported by kite twine or even rubber bands. Such an antenna is capable of delivering excellent performance and is difficult to notice from an adjoining home. The problem with such a system is that any sort of adverse weather—high winds, ice, and the like—can bring it down quickly. Birds also have difficulty in noticing the fine wire and sometimes collide with it. (In most cases, the antenna suffers more damage than the bird.)

Listeners living in apartments and condominiums, particularly high-rise buildings, have a more difficult set of problems to deal with. Often, they have no access to the ground or support structures at all. SWLs living on upper floors also have to contend with strong and unpredictable winds, which can damage an antenna structure or make it a hazard to those below. Yet such buildings are invariably constructed of steel or concrete, making an outside antenna a necessity.

Your author can sympathize with such situations, since he lives on the thirty-sixth floor of a 42-story condominium. Using another building as a support structure is not only impractical; it's impossible—the condominium building is located on the shoreline of the Hudson River, and that's all your author can see when he looks down from his apartment windows. The solution? Your author uses a 30-foot-long random wire to feed an antenna tuner. Whenever it's time for a listening session, your author opens a window next to the receiver, carefully lowers the random wire down the side of the building, and then goes SWLing. When the listening session is finished, the wire is pulled back inside and set aside until the next time. While the performance of such a system is not equal to a good random wire or dipole, the system does perform well enough to allow reception of all major international broadcasters plus a fair share of DX stations. Moreover, it is simple, safe, inexpensive, doesn't bother anyone, and allows your author to continue to enjoy his hobby despite a prohibition on outside antennas or other structures by the condominium association. Listeners in similar situations also use this approach with some interesting variations. One listener uses a fishing reel to raise and lower the antenna wire. Others obtain acceptable results by merely leaving the random wire resting on a window sill rather than lowering it.

The moral of all this is that you shouldn't despair if you live in a "problem" environment. As long as you can get *some* wire outside and feed its output to an antenna tuner, you can enjoy SWLing.

ACTIVE ANTENNAS

An *active antenna* is a physically short antenna element (often measuring less than five feet) connected to an amplifier stage. The antenna element in some active antennas is designed for mounting outside and is connected to the amplifier unit by coaxial cable. Other active antennas have the antenna element built into the amplifier unit and are designed for use indoors. Several varieties of active antennas have been introduced in recent years, and each is touted as the ultimate solution for SWLs unable to install a full-size random wire or dipole antenna. Even some SWLs who live in homes where they could erect a "normal" antenna have opted for an active antenna.

Are active antennas really as effective as their manufacturers claim? It depends upon the situation. They *can* be useful in some cases. But if you have the space to install a properly tuned random-wire or dipole antenna, you'll find those will outperform an active antenna in virtually every instance.

An active antenna is subject to the same rules as other antennas. For example, the antenna element itself must be outside any steel or concrete structure to be effective. An indoor active antenna in such situations will be just as effective—or, more accurately, *ineffective*—as any other type of antenna. Moreover, the amplifier unit will amplify atmospheric and local electrical noise as well as the signal.

The most critical part of an active antenna, and the most likely source of trouble, is the amplifier unit. The amplifier unit is, in effect, an extra receiver RF amplifier stage which precedes the actual receiver RF amplifier section. As such, it is subject to all the problems that a receiver RF amplifier is subject to, such as overloading and the spurious signals that result. One particular problem with some models is that signals from strong local AM broadcast band stations can cause spurious signals throughout lower shortwave frequencies. This is especially frustrating since AM broadcast signals are usually strongest in those urban areas where active antennas are most likely to be employed. To combat this, most active antennas include an attenuator or gain control. Unfortunately, reducing the gain of an active antenna enough to prevent overloading often makes the antenna of marginal usefulness.

Amplification of the signal presents another problem. Any amplification of received signal, whether by a receiver or an active antenna, introduces some noise into the signal. This is acceptable so long as the signal is amplified more than the noise level introduced. However, it does mean that "passive" antennas such as random wires and dipoles will give quieter signals than any active antennas. And the noise level introduced by an active antenna may vary from frequency to frequency.

Another consideration is that an amplifier unit cannot amplify what it doesn't have, and there is only so much signal that a short antenna element can "absorb" and deliver to the amplifier unit. The weak signal delivered by a short antenna element can easily be masked by amplifier unit noise and natural atmospheric noise. Even a high-performance amplifier unit cannot overcome all the limitations inherent in using a very short antenna.

The performance of active antennas varies widely among different models. An active antenna made by one well-known manufacturer was found in testing to have widely varying gain and noise at different frequencies; at some frequencies the amplifier unit delivers *no* gain and, coupled with the noise introduced, the result is a signal which is worse than if the amplifier unit were not used. Performance claims for active antennas, such as "performance equal to a 100-foot random wire," should be taken with a grain of salt. Does such a claim refer to a tuned or untuned random wire? Does it mean the active antenna's performance is equivalent to the random wire on all frequencies or on a just a few frequencies? Without such information, it's impossible to properly evaluate such manufacturer claims.

If you do decide to try an active antenna, purchase it from a dealer or manufacturer who will allow you to return it after a trial period. Some SWLs swear by active antennas, while others swear at them.

PREAMPLIFIERS AND PRESELECTORS

A *preamplifier* or *active preselector* is an external RF amplifier stage placed between an antenna and a shortwave receiver. Most also include an antenna tuning circuit in addition to an amplifier section; the amplifier section can usually be switched off and the antenna tuning unit can be used without the signal's being given extra amplification.

Many of the same comments regarding a receiver's RF amplifier section and active antennas apply to preamplifiers or preselectors. Overloading can take place from strong signals on shortwave and the AM broadcast band, although most preamplifiers and preselectors are better in this regard than active antennas. Gain can be controlled continuously, and the units can be "peaked" for best performance in a desired frequency range.

Preamplifiers and preselectors must be used carefully, since many receiver RF amplifier sections cannot handle the added output from them. Most less expensive receivers cannot be used with one; the result is a multitude of spurious signals. Excellent dynamic range, such as that found in quality receivers, is a must. Since the added gain from a preamplifier or preselector is often not necessary, many SWLs use such units primarily as antenna tuners and switch in the added gain only when they find a signal too weak to read without additional amplification.

Preamplifiers and preselectors work well with a wide variety of antennas. An outdoor random wire (or multiband dipole) connected to a preamplifier or preselector is an ideal arrangement for both SWLing or DXing. Moreover, short antennas can be used with them to make your own "active antenna," and usually at a lower cost than an active antenna system.

Preamplifiers and preselectors are often unneccessary, particularly if you can erect a tuned outdoor antenna of some type. Even then, however, such a unit can be useful in conjunction with a quality receiver in DXing situations; a preamplifier/preselector can make the difference between hearing a signal and not hearing it.

ANTENNAS FOR THE BROADCAST BAND AND LONGWAVE

Listeners interested in tuning the AM broadcast band and longwave have special antenna needs. The sheer physical size required for resonant antennas at those frequencies is a problem; a dipole for 1000 kHz (near the middle of the broadcast band) would be 468 feet long. Compounding the problem, antenna tuners, preamplifiers, and preselectors generally operate only on frequencies above 1600 kHz. Thus, specialized antenna systems have been developed for use below 1600 kHz.

The most effective antenna for such antennas is known as the *Beverage*. This is an extremely long (over 1000 feet) wire, in a straight line, which is installed only a few feet above the ground and is terminated through a resistor and extensive ground radial system. The Beverage is very directional, having a reception pattern in the direction of its terminated end which is very narrow. As you might expect, the gain of such an antenna is enormous. While impractical for most SWLs, the performance of such antennas is amazing. Listeners along the east coast of the United States have used Beverages for reception of Australian stations on the broadcast band.

Far more practical are the so-called *loop* antennas. Loop antennas come in different types and sizes, but are all compact, indoor antennas which have directional characteristics. Unlike dipoles, the directional characteristics of a good loop are pronounced. For example, it is possible to reduce or eliminate interference from stations on the same frequency. A loop has a "figure-8" reception pattern. Signals from stations located at right angles to the plane of the loop are minimized, while signals from stations along the plane of the loop are maximized. For example, if you are trying to receive a station located to the north or south of your location, a loop will reduce interference from other stations on the same frequency located to the east and west of your location. Given the great crowding found on the standard broadcast and longwave bands, this *nulling* capability is valuable.

Loops come in two basic varieties. The *air core* loop consists of a frame shaped like the letter "X" around which several turns of wire are wound. The frame is made of nonmetallic material and is from two to four feet long on each side. A *ferrite core* loop is much smaller. It consists of a bar or rod of a magnetic material known as ferrite, around which several turns of wire have been wound. A ferrite loop is enclosed inside a plastic or metal housing (the latter with openings at each end to provide a signal path to the antenna). Both types of loops can rotate, and most allow the antenna to be tilted as well (this often helps in nulling local stations). Since loops are less efficient than wire antennas, amplifier stages tuned for AM broadcast and longwave frequencies are normally used. These amplifiers are built into the loop structure and are used to tune the loop as well. The noise introduced by an amplifier is not as serious on the AM broadcast and longwave bands, since atmospheric noise is already high on such frequencies and any additional noise from the amplifier is "masked" by atmospheric noise.

The output impedance of most loops is a few hundred ohms. As such, a loop should be connected to the high-impedance input of your receiver.

If you're seriously interested in DXing the standard broadcast or longwave frequencies, a loop antenna of either type is a necessity. You can build one for yourself, using plans available from a club specializing in AM broadcast listening, or you can obtain assembled and working units from SWL equipment retailers.

AUDIO FILTERS

An *audio filter* is a device which processes the audio output of a receiver, allowing audio frequencies in certain ranges to pass unhindered to your speaker or headphones while attenuating or blocking other audio frequencies. Audio filters are poorly understood by SWLs, and are the subject of considerable hyperbole by some manufacturers. (One claims its audio filter can improve a radio's selectivity

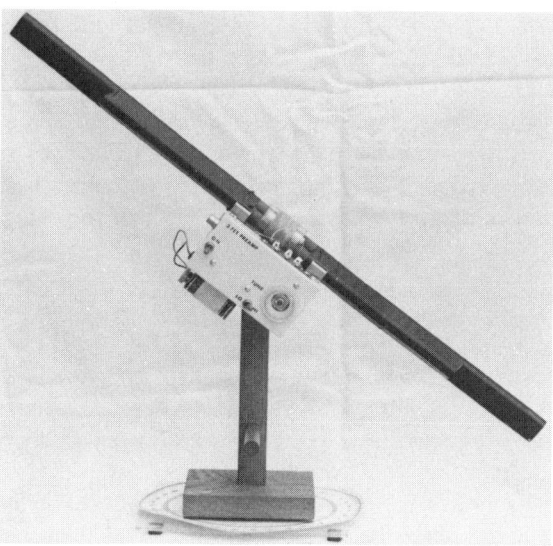

Figure 4.7 Radio West manufactures this ferrite loop antenna which can be used on frequencies from longwave to the tropical broadcasting bands.

". . . by up to 200 to 1."). The best way to appreciate the usefulness of audio filters is to think of them for what they really are—"super" tone controls.

No audio filter can make up for poor selectivity in the receiver's IF sections. If a strong signal from a nearby frequency is overpowering a desired one in the receiver's IF stages, then no audio filter will enable you to hear the desired signal. The specifications for an audio filter, such as "2 kHz selectivity at 6 dB down," refer to *audio* frequencies rather than radio frequencies. What an audio filter can do, however, is "clean up" the audio output of a receiver and make the final signal you hear more intelligible.

Audio filters are usually capable of four distinct functions. One is the *high-pass* function, which means the filter will allow all audio frequencies above a desired one to pass without attenuation but will sharply reduce or eliminate all lower frequencies. A *low-pass* function does the opposite by allowing all frequencies below a certain point to pass without attenuation. A *peak* function allows a certain range of audio frequencies to pass without attenuation but rejects all frequencies outside that range; the frequencies attenuated may be above and below a desired range. Finally, the *notch* function works in much the same manner as an IF notch filter; it removes a "slice" of the audio output of the receiver and is used primarily to reduce "whistles" resulting from heterodynes.

Audio filters are generally most effective when used for reception of SSB and CW signals, which are already "low fidelity" to begin with. The high-pass and low-pass functions are useful in reducing noise, hum, and "crud" present in a receiver's audio output. The peak function of an audio filter is especially effective for CW, since a much narrower bandwidth can be used for CW (which consists only of a single tone). The peak function is less useful with the complex audio of music and human speech. Another use for audio filters is in tailoring receiver audio to your hearing peculiarities.

Audio filters can be a useful complement to your receiver, even if they can't replace good IF selectivity. They can't solve all your selectivity problems, but they can solve some. A receiver with excellent IF selectivity used in conjunction with an audio filter is a great combination for difficult receiving situations.

HEADPHONES

You'll soon discover that many interesting signals and rare DX are only heard during odd hours, such as 3:00 a.m. your local time. If you share your living space with a family or roommates, you'll find that a good pair of headphones is the only thing that prevents you from being an outcast!

The best headphones for SWLing are not the kind you use with a stereo system. For one thing, stereo headphones are designed to work with two audio signals (the left and right channels of stereo sound) rather than the single, monaural signal produced by a shortwave receiver. Connecting stereo headphones directly to a SW receiver will result in sound being heard from just an earpiece.

Adapter plugs are available to allow stereo headphones to be used with monaural sound sources, but there is another reason for avoiding using stereo headsets— they are designed for the broadest possible audio response. That's good for listening to music, but not so good for SWLing. SW radio is inherently a low-fidelity medium, and the major emphasis is usually on clearly understanding a signal. A pair of headphones with a wide frequency response is more prone to clearly reproduce noise, hiss, heterodynes, and other sounds which decrease clarity of a signal.

Much better suited for SWLing are *communications* headphones. Not only are these designed for use with the monaural headphone jack found on SW receivers, but their frequency response is mainly limited to the audio frequencies for voice. This generally increases readability of the signal, although you won't get maximum possible fidelity during music.

An added advantage of headphones is that they allow you to concentrate better on weak or noisy signals by shutting out extraneous noises.

RADIOTELETYPE RECEIVING EQUIPMENT

Perhaps the biggest recent development in SWLing has been the wide availability of RTTY receiving equipment. There are two types of RTTY gear available; the first type is exclusively for RTTY use, while the second is composed of interfaces which turn personal computers into RTTY receiving gear.

So-called *dedicated* RTTY terminals take audio output from a receiver and convert it into a form which can be displayed on a video screen or printed out on a printer. The new generation of terminals can accept and process a wide variety of signals, such as CW and the ASCII code used by personal computers. (Radio amateurs and others use ASCII.) These terminals are extraordinarily versatile; they can handle a wide range of RTTY and CW speeds as well as RTTY frequency shifts.

A *personal computer interface* is used in conjunction with appropriate software. Software is available for most popular personal computers; the job of the interface is to take the output of the receiver and convert it into a form that can be used by the personal computer and software.

Personal computer interfaces and associated software offer an inexpensive way to get involved with RTTY reception. However, dedicated terminals give greatly improved performance and are the choice of serious RTTY listeners. The subject of RTTY reception could fill—and has filled—entire books devoted to it. RTTY reception equipment and techniques, while becoming progressively easier to use and master, are still the territory for advanced SWLs. Information and books on RTTY are available from SWL equipment dealers.

5

RADIO PROPAGATION

Propagation is a term used to describe the process by which a radio (or television) signal travels from the station's transmitter to your listening location. Understanding propagation is important for casual shortwave listening, while being able to recognize and predict unusual propagation conditions is essential for DXing.

Reception from your local AM, FM, and TV stations is via *ground wave* or *space wave*. Ground wave (sometimes called surface wave) propagation is best on lower frequencies (below 2000 kHz) and is how local AM broadcast band stations are received. As the name implies, the ground wave travels outward from the transmitting antenna over the ground. As you might expect, the ground wave weakens the farther a listener is from the transmitter antenna. But the strength of the received ground wave also depends upon the type of "ground" the ground wave travels over. Rocky soil absorbs more ground wave signal than other types of soil, for example. The least signal absorption takes place over salt water, and some amazing distances can be spanned by ground waves traveling over ocean paths. AM DXers along the east coast of the United States are familiar with this phenomenon; listeners located at the Outer Banks of North Carolina can listen to Florida AM stations at local noon, while listeners on Cape Cod can hear AM stations from the North Carolina coast. Regardless of the type of surface the ground wave travels over, the ground wave weakens as it travels away from the antenna until it is too weak to be heard above atmospheric noise. Within the area over which the ground wave travels, however, it provides steady, fade-free reception of the station.

As the frequency of a signal increases, the ground wave diminishes. At the frequencies at which TV and FM stations are found (above 54 mHz), the ground wave, for all practical purposes, does not exist. Local reception at such frequen-

Chap. 5   Radio Propagation   83

cies is provided by the space wave (sometimes known as the *direct wave*). The space wave travels through open space, much like a beam of light, from the antenna to the listener. The range of the space wave can be thought of as including everything out to the Earth's horizon as viewed from the topic of the transmitting antenna. If the horizon, as viewed from the antenna, is 100 miles distant, then the space wave can be received 100 miles distant. Actually, the space wave can be received somewhat beyond the actual horizon, since some of the space wave will be diffracted past the Earth's horizon. The space wave explains why FM and TV stations try to use tall transmitting towers or antenna locations atop mountains or tall buildings. As a general rule, reception via space wave is possible whenever the transmitting and receiving antennas can "see" each other.

Long-distance reception beyond space-wave range on frequencies from 2000 to 30000 kHz (and to a large extent on AM, FM, and TV) depends upon the Earth's *ionosphere*. The ionosphere is part of the atmosphere which extends from approximately 30 to over 600 miles above the Earth's surface. The ionosphere can *refract* back to Earth radio waves which strike it, allowing reception of the radio waves at a considerable distance from the transmitter. A signal which is refracted off the ionosphere is called the *sky wave*. Figure 5-1 is a simplified illustration of this process.

Refraction off the ionosphere is possible because of ultraviolet light radiated from the sun. The ultraviolet light *ionizes* the ionosphere (that is, the atoms of the gases constituting the ionosphere gain an extra electron) and allow refraction of signals. As you might expect, the ionosphere's ability (or inability) to refract signals depends on the level and type of solar activity.

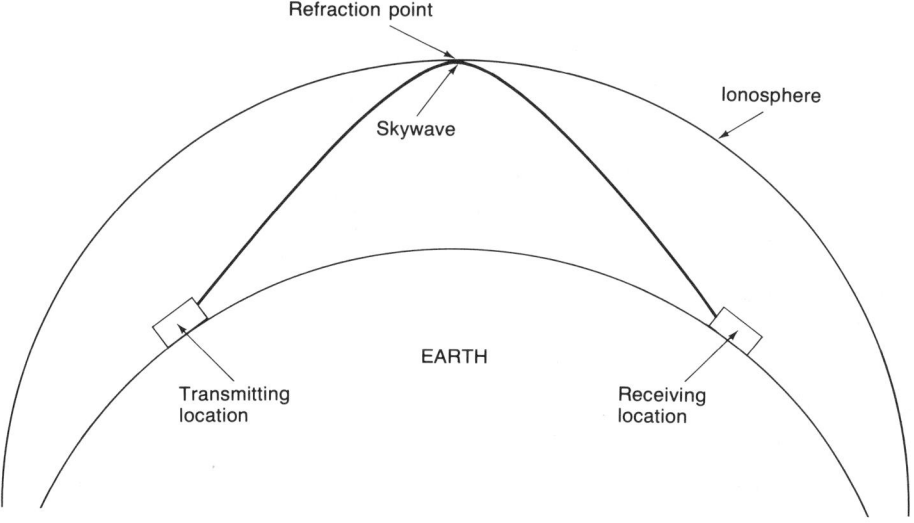

Figure 5.1 Sky wave propagation via refraction by the earth's ionosphere.

THE NATURE OF THE IONOSPHERE

The ionosphere is not a smooth, homogeneous layer of the atmosphere; rather, it is layered. Moreover, the layers are not constant. The ionization of the various layers and even their height above the Earth's surface varies with the time of day and the season of the year. Radio signals can be refracted off one or more of these different layers. However, radio signals can be *absorbed* by some layers instead of being refracted, while radio signals of different frequencies pass through the ionosphere without refraction and are lost into outer space.

The layer of the ionosphere closest to Earth is the *D-layer*. It begins at approximately 30 miles above sea level and extends upward to a little over 60 miles. Ionization in the D-layer is the lowest of any part of the ionosphere. Indeed, it is so weak that most radio signals simply pass through it to another ionospheric layer, although the D-layer usually absorbs some energy from signals and weakens them somewhat. This effect can be especially pronounced for medium- and high-frequency signals if D-layer ionization is stronger than normal. The only signals which can be refracted by the D-layer are those below approximately 300 kHz. The ionization of the D-layer is greatest around your local noon and quickly drops as sunset approaches; at night, D-layer ionization is usually negligible. During the winter months, the D-layer, for all practical purposes, ceases to exist.

The next ionospheric layer is the *E-layer*, which does play a significant role in long-distance propagation. The E-layer is found from approximately 60 to 100 miles above the Earth, although the exact altitude varies from season to season. Like the D-layer, maximum ionization occurs at local noon and diminishes as sunset approaches. Unlike the D-layer, however, the E-layer may retain enough ionization at night to affect radio propagation. The most curious aspect of the E-layer is the phenomenon known as *sporadic-E* propagation. Cloud-like patches of intense ionization can form in the E-layer and refract signals, such as the FM broadcast band and the lower TV channels (2 through 5), which normally pass through all layers of the ionosphere and into space. Propagation of such signals out to approximately 1200 miles has been observed, with refraction off several sporadic-E patches (known as *multihop*) allowing propagation over paths in excess of 2500 miles. Most typically, the range of sporadic-E is a few hundred miles. This phenomenon is most common during late May, June, and July in North America, with a smaller peak of activity often found around the winter solstice. Sporadic-E usually occurs in the morning and early afternoons, although it can unexpectedly show up at any time of day or year. Since the main impact of sporadic-E is on frequencies above 30 MHz, it will be discussed further in the section on FM and TV DXing.

For most SWLs, the *F-layer* is the most important, since it is here where most long-distance radio propagation takes place on frequencies from 2 to 30 MHz. The F-layer begins at approximately 100 miles and extends upward to over 250 miles above the Earth. During the daytime, the F-layer "splits" into two separate regions known as *F1* (which extends from 100 to 150 miles in altitude) and *F2* which begins at the end of the F1 and continues upward. The F1-layer exists in daytime and affects some radio signals, although most signals that can penetrate the E-layer can also penetrate it. At night, the F1-layer weakens and "merges" with the F2-layer to form a single F-layer. The F-layer is the last refracting layer; if a radio signal can penetrate it, the signal travels out into space and is lost. The highest frequency the F-layer can refract depends upon its ionization. Normally, the maximum frequency it can refract is between 20 to 30 MHz (and may be even lower). However, during exceptional conditions the F-layer can refract signals in excess of 50 MHz.

As mentioned earlier, the ionosphere is ionized by the sun. The level of ionization and the resulting effects on radio propagation largely depend upon solar activity. Fortunately, most solar activity can be observed, measured, and predicted with reasonable accuracy, which permits SWLs to have a good idea of what reception conditions for various frequencies will be like.

THE SUN AND RADIO RECEPTION

The sun's effects on radio reception can be predicted with reasonable accuracy by considering four factors: (1) time of day, (2) season of the year, (3) your listening location, and (4) the level of sunspot activity. This is because the sun is a stable star compared to many others in the universe (a good thing for life as we know it, as well as radio propagation). However, it's impossible to predict solar activity with total accuracy, because the sun is subject to many short-term, erratic phenomena (such as solar flares). These unpredictable events can totally disrupt normal shortwave frequencies or allow reception of DX stations which are otherwise impossible to receive.

As mentioned in the previous section on the ionosphere, ionization of the atmosphere varies with the time of day. Ionization is usually least at your local sunrise. As the sun rises, it reionizes the ionosphere. At local noon, the sun's radiation intensity is greatest and ionization is at maximum. As the sun moves toward the western horizon, the radiation intensity begins to drop, and the level of ionization starts to decline. After sunset, the level of ionization begins to fall rapidly, since no further radiation from the sun is being received. This decline continues throughout the night until sunrise, when the ionization process begins again.

You might also suspect ionization would be greater in summer than winter. If so, you're correct. The more direct rays during summer produce greater ionization than winter. And, since the seasons are reversed in the southern hemisphere, when the northern hemisphere is experiencing maximum ionization, the southern is experiencing minimum (and vice versa, of course). However, the effects of this greater ionization are probably different from what you might expect. During the winter, ionization is usually so low that the F1-layer merges with the F2-layer in daytime. This results in a single F-layer of ionization which is quite dense during the daytime. However, this single F-layer loses ionization rapidly at night. In summer, the sun's radiation is more intense due to the more direct angle of its ray. But the ionosphere is warmed by the sun more in summer, and the F-layer expands into the F1-layer and F2-layer. Since the F-layer expands, the resulting ionization density is actually *less* in summer than in winter. After sunset, though, the F-layer retains more ionization in summer than winter; this is because the ionosphere cools at night and contracts into a smaller, and therefore more densely ionized, layer.

Location also affects the ionization of the ionosphere. The sun's rays are most direct and intense at the equator, and thus the ionosphere is most highly ionized above the equator. Ionization is also intense between the Tropics of Capricorn and Cancer. Ionization decreases the closer to either the North or South Pole one goes; the ionosphere over Atlanta would be more highly ionized than the ionosphere above Montreal if measurements were taken at the same moment. At the poles themselves, some very unusual propagational conditions can be found. These will be discussed later in this chapter.

Finally, ionization varies proportionally with the level of sunspot activity observed. This is because the sun emits more ultraviolet radiation during periods when large numbers of sunspots are observed. The greater the ultraviolet radiation, the greater the ionization of the atmosphere. In fact, the number of sunspots visible is an excellent measure of ultraviolet radiation.

Sunspots appear on the sun in a series of cycles. A *sunspot cycle* is defined as the period from a sunspot minimum through a peak and down to a minimum again. Cycles take several years to complete, with 11 years being an average cycle. However, sunspot cycles have been observed as lasting from a little over seven years to over 17 years in length. Reliable records of sunspot activity and cycles have been kept since the mid-eighteenth century. At maximum, the sunspot count of some cycles has exceeded 150 (a sunspot cycle which peaked in March, 1958 had a maximum count of over 200), while the count during a cycle minimum usually drops below 10. For all recorded cycles since 1750 (as this is being written, the twenty-second such cycle is just beginning), the average maximum sunspot count is slightly over 110.

The number of sunspots on the sun changes gradually. Other solar phenomena are more abrupt. One that causes drastic effects on radio propagation is the *solar flare*. A solar flare is a sudden eruption of gas from the sun's surface which ejects a large amount of ultraviolet light, cosmic radiation, and X-rays. These travel to Earth at the speed of light and start to affect the ionosphere as soon as the flare can be visually detected. Solar flares also release large quantities of charged particles, which travel to Earth more slowly, eventually arriving from several hours to over a day after the flare itself. One consequence of a solar flare can be a *sudden ionospheric disturbance* (SID). A SID begins as soon as the flare is visible on the sun and can last any period from a few minutes to a few days. The effects of a SID can range from minor disturbances in propagation to a complete "blackout" in which no sky wave propagation is possible on SW frequencies. SIDs generally affect only those areas of the ionosphere which are in daylight when the solar flare takes place.

Another effect of a solar flair is the *ionospheric storm*. This is caused by the charged particles which arrive some time after the flare is detected. The charged particles are attracted by the Earth's magnetic field and enter the ionosphere at the North and South Poles. These charged particles are responsible for auroral displays in areas near the poles, and a brilliant auroral display is almost always accompanied by some disruption in radio communications. However, the effects of "auroral propagation" are noticeable in areas (such as the southern United States) where visual auroras are once-in-a-lifetime events.

The effects of an ionospheric storm are varied and unpredictable. For example, the F-layer may split into multiple layers or disappear, for propagation purposes, altogether. The most common effect is reduced ionization of the ionosphere and greater absorption of the signals it does refract. However, ionospheric storms of moderate intensity can actually be helpful to DXers. Since charged particles enter at the polar regions, those areas are affected first. The charged particles then work their way toward the equator. Most of the charged particles do not travel any further to the equator than the middle latitudes, however. The ionosphere over the tropical regions is thus seldom affected by ionospheric storms. The result for listeners in North America is that signals from the east or west of their locations may be severely weakened or totally missing, while signals from locations to the south are unaffected. This means a band normally crowded with powerful international broadcasters from Europe, such as 49-meters, may be free of those stations, or their signals may be significantly weaker. This allows reception of rarer domestic broadcasters in South America whose signals are normally covered by the powerhouses. Ham radio operators who operate on the higher amateur radio frequencies, such as the 144–148-MHz band, also find auroras useful. The ionization of an aurora is often so intense that it is capable of refracting radio signals which normally pass through the ionosphere. In auroras, hams use directional antennas to aim their transmitter's output north toward the aurora, which then refracts their signals back to the south. For example, two hams in New York and Minnesota could contact each other on 144 MHz by pointing their antennas north and using the aurora. Such techniques allow a 144-MHz ham station, which might normally have an effective communications range of only 100 miles, to contact other hams at distances in excess of 1000 miles. (This is a distance similar to sporadic-E propagation, and the distance is no coincidence; auroral displays happen in the E-layer.)

Unlike SIDs, the effects of an ionospheric storm are present on both the day and night sides of Earth. Higher frequencies are usually affected first, with lower frequencies experiencing disturbed conditions later and usually not as severely. (However, the effects of an ionospheric storm can be noticed on the AM broadcast band.) During an ionospheric storm, conditions gradually deteriorate and return to normal, unlike the rapid effects of a SID. And ionospheric storms may take place without a SID's taking place first. Many flares are not strong enough to cause a SID but do release substantial quantities of charged particles which do cause ionospheric storms. It won't take many hours of tuning SW before you will be able to immediately recognize when conditions are "auroral," usually by the weaker, more "watery" quality of stations to your east or west which are normally loud and reliable.

PROPAGATION PATHS

Many things happen to a radio signal in the interval (measured in millionths of a second) between when it leaves a transmitter and when it is received by you. The signal may be refracted by one *or more* layers of the ionosphere. Some of the signal's energy will be absorbed by the various layers of the ionosphere. While radio signals "normally" (if there is such a thing in radio propagation) follow the shortest route from one point to another, it is possible for a signal to be propagated along a *longer* route than the shortest one. Some parts of the signal may arrive at a receiving location before or after other components of the signal, resulting in fading or distortion. The route a radio signal takes is known as its *path*. Path means more than just the physical distance a signal takes in reference to the Earth's surface; it also refers to how it is refracted by the ionosphere. Thus, *multipath* means a signal that is received via different paths simultaneously. The result is a degraded, distorted signal.

There are several terms used to describe propagation paths you should become familiar with. One is *maximum usable frequency* (MUF). This refers to the highest frequency that can be refracted by the ionosphere along a given path. Any frequency less than or equal to the MUF will be refracted by the ionosphere, with signal strength increasing as the transmitted frequency becomes closer to the MUF. This is because signal absorption by the various layers of the ionosphere drops as the transmitted signal approaches MUF and is minimum when the signal frequency equals MUF. But if the frequency is above MUF, the signal passes through the ionosphere into space. Since MUF depends upon ionization levels in the various ionosphere layers, MUF varies almost daily, depending upon the four factors discussed earlier.

The opposite of MUF is the *lowest usable frequency* (LUF), which is defined as the lowest frequency on which an intelligible signal can be transmitted over a given path. At frequencies below the LUF, the ionosphere will absorb too much of a signal, or the atmospheric noise will be too great to permit reception of the signal that is refracted. Unlike MUF, LUF depends to a degree upon transmitter power used and the receiving equipment used; a powerful transmitter and excellent receiving gear can maintain communications at lower frequencies than could be established by lower transmitting power and more modest receivers. However, there are limits to this approach; there comes a point beyond which no practical (or even theoretical) increases in transmitter power or receiving performance will lower LUF. Like MUF, LUF varies on a daily basis.

The concepts of MUF and LUF are important to keep in mind as you tune shortwave—or any other frequency, for that matter. While technology has certainly made SWLing easier, SW reception still depends on some natural forces beyond our control. If you're trying to hear a station whose operating frequency is above the MUF for a given path, nothing will allow you to hear that station. You can try a better receiver, switch to a longer antenna, or switch on all the preamplifiers and audio filters you want—you'll still hear only noise, not the desired station. The same thing applies if you try to hear a station operating below the LUF for a given path. The only way you can count on reliable, daily reception from a station is to be within its ground or space wave coverage. Fortunately, most day-to-day variations in the MUF or LUF are minor, and it is possible to make reasonably reliable predictions of reception conditions along various paths. And as for the unpredictable events such as SIDs and ionospheric storms—well, they help keep SWLing, and especially DXing, challenging.

From previous sections of this book, you may have an idea of how MUF and LUF vary by season and time of day. In winter, the denser ionization of the F-layer means that the MUF is higher in daytime than it is in summer. However, since the total ionization is less, the night MUF is lower than it is in summer. During summer, the "expanded" F-layer is not as densely ionized during daytime as in winter, and the MUF is consequently less. However, the total ionization is greater in summer, and thus the night MUF is greater than in the winter. This also means that higher frequencies are generally useful during the daytime while lower frequencies are useful during hours of darkness.

Another important term to understand is *great circle path*. This is the path a radio signal normally takes, which is the shortest direct route between two points possible. Whether reception is possible along a great circle path depends on the frequency of the signal and the state of the ionosphere between the station and the listener. Sometimes reception is possible along a much longer path between two stations; this is known as *long-haul path* reception. The long-haul path is usually the reciprocal of the great circle path between two points.

One of the most useful tools in understanding great circle paths is a globe. Maps and atlases can be important tools for SWLs, but any flat representation of the Earth's surface is inherently distorted and can lead to misleading conclusions about where the great circle path between two locations would lie. For example, the shortest path between two points on Earth is never a straight line; it is always a *curved* line, since the Earth is round. (All true straight lines, in the geometric sense, between two points on Earth must pass *under* the Earth's surface.) The significance of this will become apparent if you spend a few minutes with a globe and string finding the great circle paths between your location and various places around the world. For example, you might be surprised to discover that the shortest route between much of North America and many nations of Asia does not lie to the west but instead is a path over the North Pole. The concepts of great circle and long-haul paths are important, because it is the MUF and LUF along those paths which allow reception. As a general rule, you'll find that great circle paths are those located mainly in daylight, while long-haul paths are located primarily in darkness.

Propagation between distant points is more complicated than simply determining the MUF or LUF, however. For example, a transmitter site may be in daylight, while the receiving site could be in darkness. Moreover, since the Earth rotates, the patterns of day and night shift and the MUF and LUF along a path are, in effect, continuously changing. This means that a given propagation path is "open" during certain times of the day but "closed" in others. It also means certain paths are open during the winter but not the summer (and vice versa) and that other paths, open during periods of high solar activity and resulting higher MUFs, are closed in years of lower solar activity. Thus, you'll often hear SWLs talk of band *openings*, which refer to those periods when propagation allows reception of stations in different areas on given frequencies. The length of an opening can vary greatly; it can range from a few minutes when sporadic-E propagation is involved to over several hours on shortwave. However, SW openings can likewise be very brief. For example, DXers in eastern North America may find the 60- and 90-meter broadcasting bands open to India around their local sunrise. Such openings to India may last as little as ten or fifteen minutes, and are present only a few days out of each year. Yet they are avidly sought by DXers, since such openings provide the only opportunities to hear those stations in eastern North America.

Not every great circle or long-haul route can support propagation at a desired frequency or to a given area, however. For example, suppose it is an early afternoon in July, and you are located in eastern North America and wish to receive stations located in Australia. At that time, however, all of Australia is in darkness, and it will be at least a couple of hours more before sunrise along eastern Australia. In this situation, it will take exceptional propagation to support an opening between the two points. Rare openings at this time can and do take place; however, most of the time propagation on any frequency between Australia and eastern North America will simply be impossible at this time in July. But a few hours later there is usually an excellent opening between the two points on frequencies from 14000 to 18000 kHz.

Discussion so far has assumed that a signal is refracted only once by the ionosphere. This is sometimes the case, but more often a signal is refracted by the ionosphere and is returned to Earth, where it "bounces" back to the ionosphere to be refracted once again back to Earth. Propagation involving more than one refraction between Earth and the ionosphere is known as *multihop* propagation (the same term used for refraction by several sporadic-E clouds). Multihop is common on most paths of more than a few thousands of miles. You'll soon learn to recognize multihop involving several refractions by rapid, rhythmic variations in signal level. On your receiver's S-meter a "long" multihop signal will "bounce" the meter rapidly.

Multihop signals are normally weaker than those from more direct propagation paths. This is because some energy is absorbed each time the signal is refracted by the ionosphere or returned from Earth. The ground is particularly absorptive of radio energy, while sea water weakens a signal much less. Thus, multihop signals which have Earth return points on the oceans are usually stronger than those which involve large land masses, such as Asia. Some of the best DX-receiving locations are those along coastlines or on islands where the last few Earth returns involve oceans. Hawaii and New Zealand are two locations where outstanding DX receptions have been reported for many years.

Multihop signals tend to be more stable on higher frequencies closer to the MUF for a given path. While the strength of such signals varies, they are usually strong enough for complete intelligibility. This is not the case as frequency drops from MUF. Lower-frequency signals may fade out altogether during multihop reception; such signals may be audible for a total of only one minute during a three-minute period. Fading patterns on low frequency signals tend to be "slower" than those found on higher frequencies; the rise and fall in signal strength is often more gradual than the abrupt changes in strength that normally characterize fading of higher frequencies.

An interesting effect is noticed when a signal path crosses through either polar region. Each pole is surrounded by an *auroral zone*. This is the region in the E-layer where auroral activity takes place, and one surrounds each pole. Even when no visible aurora is present, there is usually enough irregular ionization in these regions to disturb radio signals passing through it. This usually is manifested by a rapid, irregular fluctuation of the signal known as *flutter*; signals may fluctuate in strength at a rate in excess of 100 variations per minute. The fluctuations may be so rapid that voice communications, whether AM or SSB, can be made unreadable, and CW is the only reliable method of communications. Listeners in North America quickly become familiar with "auroral flutter" by its effects on signals from Asia (including the Asiatic USSR) arriving over the North Pole.

Ionospheric refraction is also more complex than assumed so far. Suppose that both the F1-layer and F2-layer are present in the ionosphere. A signal could be (and often is) refracted by both layers. Since the F2-layer is higher than the F1-layer, any signal refracted off it would have to travel further from the transmitter site to the refraction point and finally to the receiving site. If the same signal were also refracted off the F1-layer, the signal from the F1-layer would arrive at the receiving site *ahead* of the same signal refracted via the F2-layer. When the signals are processed by the receiver, the two identical signals arriving within microseconds of each other "confuse" the receiver's mixer stage. The result is distorted audio; in some cases, the two signals cancel each other out almost entirely. Making the situation even more complex is the fact that the E-layer can also refract some signal, and multihop reception may be involved. Listeners to the

AM broadcast band at night may find a similar phenomenon taking place when tuning stations are located a few hundred miles distant. In such situations, the station's signal may arrive simultaneously via ground wave and sky wave, with the ground wave arriving microseconds ahead of the sky wave and producing distorted audio from the receiver.

It is also possible for the same signal to arrive at a receiving site by different paths. This is known as *multipath* reception. For example, the great circle route and long-haul path may both be open at the same time. Or a signal may be propagated along two different paths by different sporadic-E clouds. Sometimes multihop reception will be referred to as *multipath* reception.

The ionosphere is obviously not static. Some effects, such as the fade-in and fade-out of various stations at local sunrise and sunset, are gradual and easy to predict. But abrupt changes can take place in seconds, and these are particularly noticeable when one is receiving low-powered stations or listening along difficult paths. Every active SWL can recall cases when a signal abruptly faded into the noise and was lost or, more happily, when a weak, barely audible signal suddenly increased in strength to full readability. In fact, these conditions are common with sporadic-E propagation above 30 MHz or during ionospheric storms on frequencies below 5000 kHz. Even a normally reliable path can undergo brief, sporadic disruption, and this is what causes fading. Propagation can vary greatly over a range of only a few kHz, especially in terms of the arrival time of a signal. You'll recall from Chapter 3 that this is why ECSSB reception is often used in difficult receiving situations. It also explains why FM, with its wide bandwidth, is used primarily in situations where the space wave is the main propagation method. When FM broadcast stations are received via sporadic-E, it is not uncommon for such signals to be badly distorted, since propagation through the range of the signal's bandwidth is irregular.

REFRACTION ANGLES AND SKIP ZONES

One thing that puzzles many new SWLs is how stations more distant from one's location can be heard well while stations located closer are heard only weakly or not at all. This is because of the phenomenon known as the *skip zone*. This is the area beyond the coverage of a station's ground or space wave but short of the area in which the sky wave returns to Earth. The station's signal literally "skips" over SWLs located in the skip zone. Figure 5-2 illustrates how a skip zone operates. Although only one refraction is illustrated in Figure 5-2, skip zones are also found during multihop reception. Other skip zones are found "under" each ionospheric refraction point.

Skip zones are not as neatly static as Figure 5-2 suggests. Even for the same station on the same frequency, there is often at least one or two periods per day when areas that can normally receive the station find themselves in a skip zone. These periods are usually around local sunrise and sunset, when reception condi-

Refraction Angles and Skip Zones 93

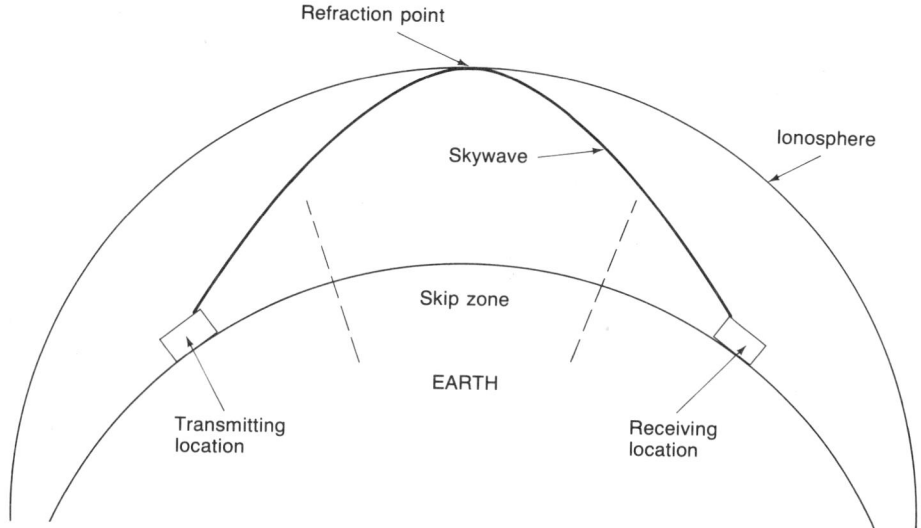

Figure 5.2 Although located closer to the transmitting station, listeners in the skip zone will be unable to hear the station because the signal is refracted overhead.

tions are changing from night to day and back again. For example, listeners in eastern North America can hear WWV in Colorado on 15000 kHz throughout the day. But eastern North America finds itself in a skip zone for WWV around sunset, and other stations can be heard during that period. For example, WWVH in Hawaii often is stronger than WWV at that time.

The distance covered by sky wave propagation depends upon the *angle of refraction*, sometimes known as the *critical angle*. The angle is defined in reference to the Earth, and the lower the angle (as measured in degrees), the longer the distance covered by the sky wave. Higher-frequency signals have a lower angle of radiation than lower-frequency signals. Figure 5-3 shows various angles of refraction and the distance covered. Note that the distance covered depends on the altitude where the refraction takes place in addition to the angle of refraction; a refraction off the F2-layer covers more distance than one off the F1-layer. The distance covered by a refraction also increases as the frequency refracted approaches the MUF. The greatest distance covered by sky wave propagation and the longest skip zone are the product of a low angle of radiation and a refraction off the F2-layer near the MUF.

A low angle of radiation produces stronger signals over multihop paths than one which has a higher angle of radiation. This is because a low angle of radiation signal covers more distance per refraction and requires fewer refractions, and consequently loses less strength, to cover a given distance.

Many stations use transmitting antennas designed to concentrate energy into a low angle of radiation. You'll also recall from Chapter 4 that vertical antennas give better reception of signals arriving at a low angle. In addition, the ability of

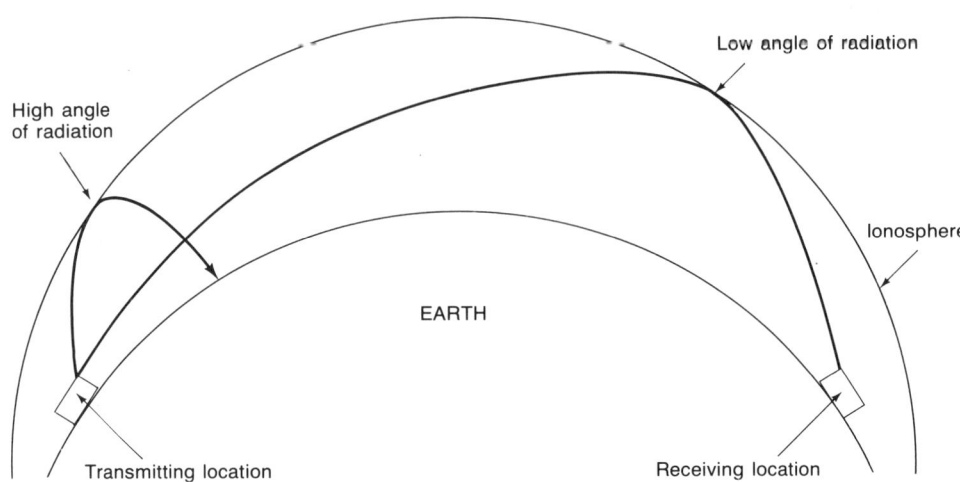

Figure 5.3 The lower the angle of refraction, the more distant a signal may be propagated.

an antenna to receive or transmit signals at low angles increases as antenna height is increased. Moreover, a clear horizon helps, since mountains and high hills can absorb signal energy arriving at a low angle. (You probably won't be too surprised to learn that many top DXers have selected homesites located on hilltops.)

RECEPTION PATTERNS

Many new SWLs are puzzled to discover that various frequencies are useful for reception at times other than those they would at first seem suited for. For example, the 19-meter international broadcasting band commonly permits reception of signals from stations to the west of one's location until very late at night. The 49-meter and 60-meter broadcasting bands are normally thought of as night frequencies, yet they usually permit reception of stations to the east of one's location beginning in early afternoon and reception of stations located to the west up until a couple of hours after local sunrise. And the 31-meter and 25-meter bands are often open to some part of the world around the clock. You'll quickly notice many other exceptions to the broad guidelines for reception times given for various frequency ranges back in Chapter 2.

As mentioned previously, the ionosphere is extremely dynamic. The state of the ionosphere along a given propagation path is constantly changing. And since most propagation paths involve multiple refractions off the ionosphere, the effects of ionospheric changes along a path can be (and often are) extremely varied. For example, the ionization level of the ionosphere at certain refraction points may be dropping while it is increasing at others. There may be only a single F-layer along parts of a path, while the remainder of it has both the Fl-layer and F2-layer. The D-layer and E-layer may be present at certain refraction points but not others.

Perhaps the most dramatic effects on propagation, particularly on lower frequencies, take place at sunset and sunrise at the transmitting site. As sunset approaches, the D-layer and E-layer near a transmitting location weaken rapidly or disappear. This allows signals to reach the F-layer (or layers) with less attenuation. Since low-frequency signals are especially affected by these layers, propagation becomes much easier on such frequencies at sunset. An even more dramatic effect takes place at sunrise at a transmitter site. The D-layer and E-layer rapidly reform within minutes after sunrise, and ionization and absorption of low-frequency signals quickly returns to daytime highs.

For SWLs, this means that sunset at transmitter sites to the west of one's location produces an increase in strength of low-frequency signals. The reverse happens when sunrise occurs at transmitter sites to the east of a listening site; there is usually a quick fade-out of low-frequency signals. Listeners along the east coast of North America can often "follow" sunrise moving across Africa on 60-meters after 0500 UTC as one station after another rapidly fades into the noise with their local sunrise. Similar effects, although not as pronounced, are found on higher frequencies.

Sunrise and sunset at listening sites are not as dramatic, because the key factor is the state of the ionosphere at various refraction points along the path. This means that low-frequency signals from the east can fade in long before sunset at a receiving site, while high-frequency signals from the west can remain audible long into the evening or night. Conversely, high-frequency signals from the east can be heard before sunrise at a listening location while low-frequency signals from the west are still heard an hour or two after the sun is up.

Advanced DXers take advantage of a technique known as *gray line propagation*. The gray line is the terminator (twilight or dawn) that separates daylight and darkness areas of Earth. Gray line propagation takes place when both the transmitting and receiving sites are along the terminator. At any listening location, gray line propagation is possible for roughly two hours per day at a half hour before and after local sunrise and sunset. Transmitting locations located along the terminator at the same time can then be received, and often the only opportunity to receive these stations takes place during gray line openings. The openings to India on 60-meters and 90-meters from eastern North America mentioned earlier in this chapter take place via gray line propagation.

With experience, you will soon become familiar with various reception patterns and what can—and cannot—be heard from your location. Experienced SWLs often have "beacon" stations for various bands to give a quick indication as to reception conditions. Major international broadcast stations are usually not good beacons, since their powerful transmitters and efficient antennas can deliver strong signals even in poor reception conditions. Far better are domestic broadcasters or standard time and frequency stations. Most DXers have a list of beacon stations around the world for different times and frequencies. If a beacon station cannot be heard, it is unlikely that other stations from that area in the same frequency range can be heard. However, if the beacon's signal is stronger than usual, conditions are above normal, and DX stations from the same area and frequency range may be heard.

PROPAGATION FORECASTING

Most major international broadcasters employ at least one specialist whose responsibility is forecasting propagation conditions along with suitable frequencies and times for paths to the station's intended target audiences. Thus, if your primary SWLing interest lies in receiving major international broadcasters, there's no need to worry about the best times and frequencies for reception; those have already been selected for you. However, if you're more interested in DX targets, simple propagation forecasting techniques will greatly improve your chances of hearing such stations. While it's difficult to predict such events as SIDs, other solar phenomena are more regular and predictable in their behavior.

One key in propagation forecasting is the sun's rotation period, which is approximately 27.5 days. For example, suppose that a large group of sunspots are present on the face of the sun, indicating a high degree of ultraviolet radiation and consequent ionization of the ionosphere. As the sun rotates, the group of sunspots moves away from Earth's view and the level of ultraviolet radiation also drops. But after 27 days, the sunspot group again rotates so that we can see it from Earth, and ultraviolet radiation increases. This means that solar effects on propagation tend to repeat in 27- or 28-day cycles. Of course, the exact conditions seldom repeat, since the sun is a dynamic body. But if you keep records of propagation conditions for a given day, and make appropriate adjustment for seasonal changes such as time of daylight and angle of solar radiation, then it's possible to predict conditions 27 days in advance with accuracy as high as 80 to 90 percent. Sometimes conditions will be better (the sunspot group gets larger) while conditions may be worse (the sunspot group shrinks). But you'll find the 27-day solar rotation cycle to be a good general guide to propagation conditions. This is particularly true during years of low solar activity and few sunspots, since the sun appears to be more stable then.

Sometimes the effects of solar flares repeat every 27 days. A weak flare which is just beginning when it rotates out of view may not deliver its full impact until it reappears. A flare spotted during a previous rotation can also have a diminished effect when it again rotates into view.

A daily indication of solar activity is provided by radio stations WWV and WWVH at 18 minutes past the hour on each of their operating frequencies. At this time, WWV/WWVH broadcasts a prediction of Earth's geomagnetic activity along with the *solar flux* and *K-index* readings. The geomagnetic activity report is fairly straightforward and uses terms such as "quiet," "unsettled," or "active," or indicates that an ionospheric storm is in progress. These reports indicate how active the ionosphere is. An "active" or "unsettled" report means that propaga-

Propagation Forecasting 97

tion conditions may be rapidly shifting and that signals may be affected adversely; this is especially true for signals traveling along paths between points in higher latitudes. An ionospheric storm means that the "auroral conditions" mentioned previously are in progress. By contrast, a "quiet" report usually means that most propagation paths are not significantly disturbed by ionospheric events.

Even if the paths are not disturbed, this does not indicate what the MUF or LUF along a path might be. One measure of the intensity of solar radiation and resulting ionization is the solar flux measurement. This is a measurement of radio noise "transmitted" by the sun on approximately 2800 MHz and correlates closely with ultraviolet radiation from the sun. The higher the solar flux reading, the higher the MUF will be on most paths.

The K-index is a measure of activity in the Earth's magnetic field as measured at Boulder, Colorado. A major factor influencing the K-index is charged particles from the sun. A small K-index number indicates good propagation between high latitude points and low ionospheric absorption, while a large K-index reading indicates the possibility of auroral conditions and high absorption along paths in higher latitudes.

While WWV/WWVH broadcasts propagation information each hour, it is not updated that regularly. The solar flux value is the previous day's reading; however, solar flux seldom varies significantly in a day. The K-index is updated each three hours and is usually no more than six hours old when given. In fact, it is possible to get the latest information via telephone; the number is (303) 497-3235. (It is a toll call, and collect calls are not accepted.)

How would the WWV/WWVH information be interpreted? The best conditions for reception would be a forecast for quiet geomagnetic activity with a low K-index number (no greater than 2) and a high solar flux reading (such as 90 or above). Similar conditions but with a lower solar flux number means that the MUF may be lower and thus high frequencies may not be as useful; however, low-frequency DX may be excellent. A geomagnetic activity report of "unsettled" or "active" but with a low K-index number and high solar flux indicates average or normal reception conditions. If the K-index is high (up to 4) or the solar flux is low, conditions will be below normal. If an ionospheric storm is in progress, or if the K-index is over 5, reception conditions will be disturbed.

Standard time and frequency stations in other part of the world, such as Japan's JJY, also give propagation forecasts. Major international broadcasters with programs for SWLs, such as Radio Canada International, also give propagation forecasts.

Several tools to assist propagation forecasting are available. One useful example is "The DX Edge," a device similar to a slide rule. It consists of a world map with clear plastic overlays representing the day and night patterns for all twelve months of the year. This allows instant determination of which areas of the world are in daylight or darkness and where the gray-line terminator lies. "The DX Edge" is available from several SWL equipment suppliers or from Xantek,

Inc., P. O. Box 834, Madison Square Station, New York, New York, 10159. In addition, several propagation prediction software packages are available for various popular personal computers. These programs take input of the date and WWV/WWVH figures for the solar flux to produce predictions of MUF and LUF along paths between various points.

6

MAJOR INTERNATIONAL SHORTWAVE BROADCASTERS

If you were to ask experienced SWLs which was the first station they heard on shortwave, the answers would almost certainly be something like "the BBC," "Radio Nederland," "Radio Moscow," or "HCJB." These are just a few of the well-known international shortwave broadcasters who beam powerful signals to North America each day. It doesn't take a great deal of effort to hear such stations; a quick scan of the 49-meter through 19-meter broadcasting bands between 0100 to 0500 UTC will let you hear dozens of different countries broadcasting in English.

Some SWLs who start off by tuning international broadcasters move on to more specialized interests, such as pure DXing or utility listening. But many stick with listening to international broadcasters for various reasons. The signals from international broadcasters are the only ones on shortwave which could be described as "easy listening"; the high power and optimized propagation paths reduce the detrimental effects of the ionosphere. Moreover, the programs themselves are intended for listeners in other countries, and some attempts are usually made to take the target audience into account in determining content. And often the news broadcasts of international stations are the only way many North Americans can learn about domestic events and issues in most foreign countries.

However, as mentioned in Chapter 1, many international stations are little more than public relations agents for the nations which fund them. This tends to be true for stations operated by countries across the political spectrum; the Voice of America may not be as strident as Radio Moscow, but the purpose of both is to create a favorable image for the operating nation. This notion hasn't been lost on emerging nations; it seems as if one of the first things many developing nations do after building a major international airport is to set up an international broadcasting service. You'll also find that several international broadcasters are operated by evangelical Christian organizations, and often reflect little or nothing of the countries they operate from. For example, HCJB in Quito, Ecuador is often the first station new SWLs hear from South America. (It may also be the first station new SWLs ever hear on SW, as was the case with your author.) However, you won't learn a great deal about the culture, politics, and lifestyles of Ecuador from HCJB, because that's not its prime purpose—evangelism is.

It was also mentioned in Chapter 1 that much international broadcasting tends to be dull. This is admittedly a subjective observation, since what one listener might find fascinating is what another might consider boring. Yet many people will consider most of international programming a "yawn." Some stations seem indifferent as to what North American listeners might be interested in or how to package information in such a way that might catch the attention of listeners. However, careful listening, combined with independent sources (books, scholarly journals, and the like) of information about a country, can make international broadcasting a prime source of material on foreign politics and cultural affairs.

One disconcerting habit of international broadcasters is their habit of switching frequencies and times of broadcasts several times per year. Most stations change frequencies four times per year, usually on the first Sunday in March, May, September, and November. The rationale for this is good; the new frequencies are selected for best propagation in light of seasonal changes and expected MUFs. However, it does mean that you may not be able to hear a station on the same frequency a week later. It also means that this book will not include frequencies for international broadcasters (with the exception of a few "traditional' frequencies, which haven't changed for years). To keep up to date with the latest times and frequencies for these stations, you can write the stations directly and ask to be placed on their mailing list for current schedules. SWL clubs and commercial publications also carry the latest schedules, and the stations themselves announce new times and frequencies well in advance.

The rest of this chapter will be devoted to profiles of major international broadcasters; each profile will include the station's mailing address for schedules and reception reports. For a more complete listing of international broadcasters and their addresses, consult the latest edition of the *World Radio TV Handbook*.

THE BRITISH BROADCASTING CORPORATION (BBC)

The BBC epitomizes international SW broadcasting for many people. This is hardly surprising. For program quality, news accuracy, and worldwide coverage, the BBC is in a class by itself. While no one at the BBC has come out and explicitly said so, the BBC seems to perceive its programming as a service to all the people of the world. It comes very close to that.

The British Broadcasting Corporation (BBC)

The BBC is an independent public corporation supported primarily by license fees paid to use radio and television receivers in the United Kingdom. It is overseen by a board of governors composed of noted and respected persons in British public life with daily management in the hands of career broadcasting professionals. The arrangement seems to work well, since the BBC is perhaps freer of governmental influence than any other publicly financed broadcasting organization. In addition to its shortwave services, the BBC also operates four radio networks on AM, numerous local AM and FM stations, and two television networks (known as BBC1 and BBC2) in Britain. There are also regional broadcasting services, such as BBC Radio Scotland for the other components of the United Kingdom. The service that SWLs are familiar with is known as the BBC World Service and it operates on various frequencies 24 hours a day.

For years the BBC had a reputation as the "stuffy old dowager" of international broadcasters, and was known as "Auntie BBC" even to its fans. Things have changed greatly in recent years, however. While there are still some traces of the previous "Oxbridge mentality," a new generation of BBC staff realizes that their audience includes more than just the House of Lords. The result has been more lively, innovative, and experimental programs without resorting to "schlock."

Incidentally, you'll find many BBC programs to be brief. Don't expect hour-long segments devoted to a single program; often, a BBC program may last as little as five or ten minutes. (This is often the case with other international broadcasters.) While the time allocated to each program may be short, you'll find that the programs don't waste time or words; they are succinct without being skimpy.

What are you likely to hear on the BBC? One staple is something you seldom hear on North American radio—drama. If you've never listened to people "act" with just their voices, you're in for a treat. You'll hear presentations of classic works, such as *Les Miserables* or *Treasure Island* as well as the efforts of more contemporary playwrights and writers. Other frequent presentations are programs devoted to a single major topic or theme, such as dyslexia or hunger. These are different from the "investigative" pieces run by American television; a more accurate term for them would be "research" programs. The BBC emphasizes a thorough digging into the facts of a given situation and a balanced presentation of differing opinions rather than an emotional approach designed to get ratings. The BBC, as perhaps the most independent of all government-financed broadcasters, is not afraid to tackle sensitive subjects even at the risk of offending the party occupying number 10 Downing Street. Many of these deal with domestic issues (such as the 1984 coal miner strike) and will not have much relevance for many overseas listeners. However, the BBC does probe subjects, such as deployment of cruise missles in Britain, that will be of interest to SWLs worldwide.

Perhaps the best-known feature of the BBC is its news broadcasts. While objectivity in news presentation is difficult, if not impossible, to objectively judge, the BBC's efforts at being impartial, fair, and balanced are recognized worldwide. In fact, the BBC may well be the most trusted news source in the world today. It is not uncommon in many parts of the world for listeners to tune to the BBC to find out what is happening in their own country. In fact, many Americans living or traveling abroad turn to the BBC instead of the Voice of America to find out what is going on in the world. You'll find BBC news summaries on the hour.

The hourly news programs are supplemented frequently by news analysis features and summaries of the British press and other radio stations around the world. (Unfortunately, the British press summaries don't capture the delightful wackiness of the numerous daily tabloids and their obsession with sex and drunken rampages by the famous.) There are also news programs devoted to British news and events. One program your author finds fascinating is "Letter from America," which offers a British perspective on American events and society.

Music is another popular subject for BBC programs. In years past, the BBC stuck to opera, classical, and symphonic works with only a passing acknowledgment of more current trends in popular music. In the early 1960s, one would have never guessed that a major revolution in popular music was beginning in Britain by listening to the BBC and its assortment of familiar standbys such as band leader Victor Sylvester. There is still much classical programming, but the BBC today also features contemporary British rock stars, jazz, and features on artists of worldwide popularity (such as Frank Sinatra). BBC music programs consist of more than a disc jockey playing records; frequently, the artists themselves are present to talk about their work. One indication of how the BBC has changed over the years is the fact that John Peel, a well-known disc jockey for the offshore "pirate" radio station Radio London during the mid-1960s, now has a rock music program on the BBC.

Programs are just as varied as the audience it serves, and range from poetry readings to features for merchant seamen at sea. There are serials, often with a science fiction theme, such as "The Hitchhiker's Guide to the Galaxy" and "Space Force." And such futuristic fare might be immediately followed by a more mundane farming report.

While only English-language programs have been mentioned so far, the BBC transmits in over 35 other languages ranging from Arabic to Indonesian to Urdu. You won't find these programs of much interest unless you understand the language, of course, but they are an important part of the BBC's reputation and following worldwide. Those whose interests lean more toward DXing than SWLing find it valuable to be able to quickly recognize the BBC in other languages to prevent spending time and effort in trying to identify a "rare" station that turns out to be common.

The BBC operates four large transmitter sites in Britain itself which use transmitters of up to 500 KW of power. To improve reception in different parts of the world, the BBC has established relay stations at locations such as Ascension Island, Cyprus, Oman, Singapore, and Berlin, and it shares a relay station on the

Caribbean island of Antigua with the West German broadcaster Deutsche Welle. In addition, the BBC World Service is relayed via the Voice of America at its transmitter sites at Delano, California and Greenville, North Carolina and by Radio Canada International from its Sackville, New Brunswick station. (In turn, the BBC makes its British transmitter facilities available to those two broadcasters.) However, unless you have a copy of the latest BBC transmission schedule or catch the station when it signs on, you won't be able to tell which transmitter site you are listening to, since all identifications are given as "BBC World Service" or "This is London" once a transmission is in progress.

Like most other international broadcasters, the BBC uses an *interval signal* to identify itself before programming actually begins. An interval signal is a distinctive sound or musical piece repeated before a station signs on; this helps SWLs find and identify the station's signal. The BBC uses the musical notes "B-B-C" repeated on a tonal scale. You can also hear the bells of London's famous Big Ben between several BBC programs.

The results of such efforts are powerful, reliable signals from the BBC. A quick scan across any international broadcasting band will almost certainly produce a frequency or two in use by the BBC World Service. Like other international broadcasters, the BBC changes the frequencies it uses several times per year. However, there are a few frequencies which the BBC has used for several years without interruption. Among these are 9410, 12040, and 15070 kHz; the latter is a good choice for daytime reception in North America.

The latest transmission schedule is available from the BBC World Service Publicity, P. O. Box 76, Bush House, London WC2B 4PH, Great Britain. If you become a fan of the BBC, you might want to subscribe to a monthly program journal known as *London Calling*. This is a bit like *TV Guide*, listing programs by time and day along with descriptions of each program. A sample copy of *London Calling* and subscription information can be obtained when you request a program schedule.

The BBC is flooded with reception reports, and as a result sends out a nonspecific "thank you for writing" card in reply. However, the BBC apparently reads all listener mail, and especially values comments and suggestions concerning their programs. Reports and letters can be sent to BBC World Service, Bush House, Strand, London, WC2B 4PH, Great Britain. By the way, the BBC operates a "World Information Centre and Shop" at Bush House in London; if you're ever in London, you can stop by and obtain BBC publications and souvenirs there.

RADIO MOSCOW

Very much the opposite of the BBC is Radio Moscow; it is purely an outlet to promote Soviet views and objectives in other nations. While only the most hopelessly naive listeners in Western nations could possibly take Radio Moscow at face value, such is not the case in many Third World nations where it has more listenership and influence. Part of this is due to the enormous resources Radio Moscow pours into broadcasting; it currently broadcasts in 60 different languages. (This count does not include languages spoken in the USSR itself, such as Russian, Ukranian, Lithuanian, and so forth.) If the only language you know is Assamese, you can't get the news from the BBC or the Voice of America. However, you can tune to Radio Moscow's broadcasts in that language.

Most international broadcasters use a maximum of only two or three frequencies at a time for programs to a specific target area. By contrast, Radio Moscow often uses several frequencies within the same international broadcasting band and uses every broadcasting band that might be open to a target area during a given period. It's not uncommon for Radio Moscow to use as many as twenty frequencies at once for its broadcasts to North America. For example, during one recent winter period Radio Moscow used seven different frequencies in the 49-meter band alone for its North American service. As you might expect, transmitter powers, while not precisely known, are obviously quite high. Radio Moscow also has transmitter locations at many different locations across the vast expanse of the Soviet Union, allowing it to find satisfactory propagation paths to almost any part of the world around the clock. The result is that Radio Moscow is very difficult *not* to hear, regardless of where you may be in the world.

Possessing such resources, Radio Moscow has not relied much on overseas relay stations. For several years, the only relay they used was Radio Sofia, Bulgaria's 9755-kHz channel for their North American Service. More recently, however, Radio Moscow has established a relay base in Cuba for programs to North and South America as well as for relays of domestic Soviet radio for Soviet personnel in the Americas.

By the way, Radio Moscow has never shown much regard for international agreements regarding broadcasting. Back in Chapter 2, it was mentioned that 7100–7300 kHz is assigned to broadcasting in Europe and Asia, while it is reserved for amateur radio in North and South America. The rules of various World Administrative Radio Conferences and the International Telecommunications Union (ITU), of which the USSR is a member and signatory to their agreements, prohibit nations in Europe and Asia from broadcasting to North and South America on any frequency in the 7100–7300 kHz range. This hasn't deterred Radio Moscow one bit. Tune across that range any evening, especially in winter, and you'll find several frequencies being used there by Radio Moscow, all identifying themselves as "the North American service of Radio Moscow." Likewise, Radio Moscow has refused to supply current data to the ITU regarding its transmitter locations and powers as required by ITU agreements. Some of the information on file with the ITU regarding the USSR dates from shortly after World War II.

In addition to its services beamed to specific geographic areas such as North America, Radio Moscow also operates its own "World Service" on a 24-hour schedule. Over a dozen different frequencies may be used at once from transmitter sites scattered across the USSR. Many of the same programs used in the specific geographic services are also used by the World Service. Radio Moscow has a similar service in French which, at the present, broadcasts seven hours per day primarily for Europe and Africa.

Most SWLs will find Radio Moscow dull. The programs tend to move along in slow motion with a plodding, methodical quality about them. The delivery of some of the announcers is such that you sometimes get the impression you're listening to one of those recordings used to announce incorrect or discontinued telephone numbers. And you won't hear any hint of bourgeois influences on the music played.

What kind of programs will you hear? There is news, of course, and commentaries on the news and world events. A familiar feature is "Moscow Mailbag," hosted by the durable Joe Adamov. Perhaps more Americans are familiar with Adamov's voice than that of any other Soviet citizen past or present. His forte is answering listener letters and questions, which can be somewhat hostile, in a smooth and polished style. The political content of his answers always follows the current Politburo line, but his style is strictly his own; sometimes he is serious, while at other times he seems genuinely offended that an "insulting" question was even asked. To other questions, he seems amused that it was even asked; one can almost see him shaking his head in disbelief that a listener could be so ill-informed. One may disagree sharply with his politics, but most listeners would agree that Adamov is an effective spokesman for Soviet interests. (Apparently Soviet leaders feel the same, since in recent years Adamov has appeared on American television several times to field questions from newspeople.)

Some letter writers to Radio Moscow have been surprised to find phrases or sections of their letters taken out of context when read over Radio Moscow. One memorable incident a few years ago involved an American SWL who, in a reception report, remarked about QRM to Radio Moscow from hams operating in the 7100–7300-kHz range (where, you'll recall, they are supposed to be and Radio Moscow isn't). The SWL in question used the term "jamming" to describe the QRM, resulting in Radio Moscow claiming that the Americans were jamming Radio Moscow's broadcasts to North America!

Other regular Radio Moscow features might be termed "slices of Soviet life." These involve interviews (with English translation when needed) with persons of various occupations living throughout the USSR. These programs are relentlessly upbeat; the only problems are either natural (the climate, and the like) or the sheer enormity of the task (such as getting something built in record time). All problems, naturally, are solved through the resolve of the people and the principles of socialism. It's unlikely that you'll find these programs enlightening unless you don't already suspect that it is cold in Siberia during winter or that Soviet children misbehave from time to time.

One recent feature by Radio Moscow is known as "Radio Bridge." This involves listeners sending in a tape to Radio Moscow and Soviet citizens recording their replies to the submitted questions; the program is formatted so that the two sides "converse" with each other.

Radio Moscow actively solicits reception reports and sends out full data QSL cards in reply. QSLs are sent out under the direction of Mrs. Eugenia Stepanova, who has handled the task for years. QSL card designs are changed many times each year. One interesting oddity about Radio Moscow QSLs is that frequencies are specified only to two significant digits. This means a frequency such as 9615 kHz will be indicated on the QSL as "9.61 MHz." Radio Moscow also announces frequencies in the same manner, so don't assume that something is wrong with your receiver's frequency readout if you find it indicating a frequency 5 kHz higher than that announced by Radio Moscow.

Radio Moscow also has been known to indicate the transmitter site used for a specific frequency on a QSL card upon request. Some SWLs like to have this indicated, because SWL hobby standards consider the different republics of the USSR as separate "countries" and each individual transmitter site as a separate station. However, Radio Moscow uses outdated ITU data to indicate which transmitter site was used, and as a result there is usually some doubt as to whether the site indicated is the actual location of the transmitter reported. Radio Moscow's willingness to indicate the transmitter site varies over time, although it is often closely correlated to the state of American-Soviet relations. Radio Moscow also sends a pennant to regular (that is, several reports over many months) reporters.

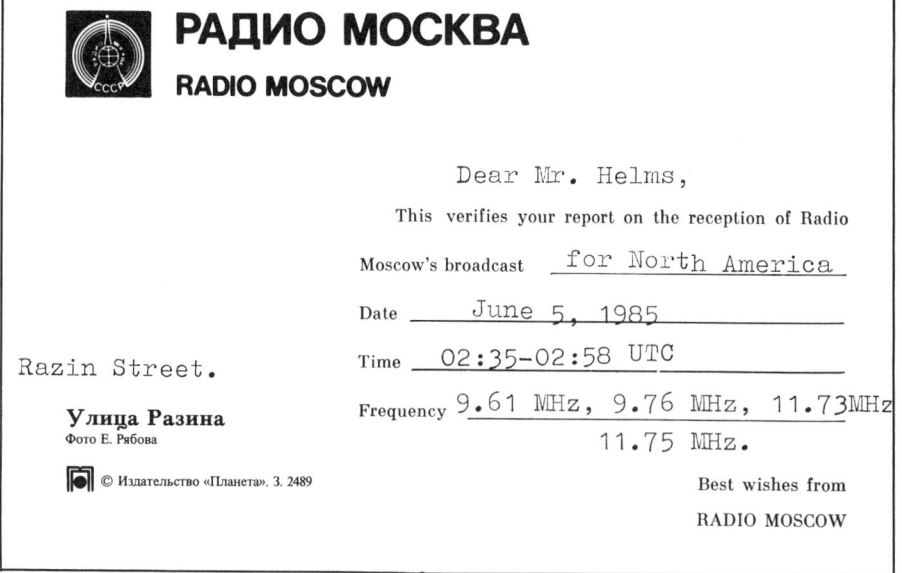

Figure 6.1 QSL card from Radio Moscow. Note how frequencies are rounded.

A single report to Radio Moscow is often sufficient to get you on their mailing list for program schedules for the next few years.

Radio Moscow's address is simple enough: Radio Moscow, Moscow, USSR. It helps to include which service, such as North American or World, you're reporting reception of in the address or on the envelope. Radio Moscow's studio address in Moscow is Pyatnitskaya Ulitsa 25; they have been known to receive American visitors with an advance appointment.

There are other international broadcasters in the USSR, although most apparently share facilities with Radio Moscow. (In fact, some of these "different" stations can be heard signing on immediately after the end of a Radio Moscow transmission with no break or change in signal strength.) One that appears to be nothing more than a separate department of Radio Moscow is known as Radio Station "Peace and Progress," which bills itself as "the voice of Soviet public opinion." Its programs are usually commentaries on the news and world events, and often they have a harder anti-American edge than those on Radio Moscow. Many of the announcers sound very much like those heard on Radio Moscow (with the notable exception of Joe Adamov). Radio Station "Peace and Progress" transmits in half-hour segments in ten languages; in addition to such familiar languages as Chinese, English, German, and Spanish, it also uses Creole and Guarani. Interestingly, the only part of the world it doesn't transmit to is North America. Despite this, Radio Station "Peace and Progress" can often be heard well during its transmissions to Asia and Africa, and it will verify reports from North American listeners. Their address is Radio Station "Peace and Progress," Moscow, USSR.

Other international broadcasters from the USSR "transmit" from the various republics making up the USSR. One is Radio Kiev, which broadcasts in English and Ukranian to North America and often uses some of the same frequencies as Radio Moscow. Their address is Radio Kiev, Radio Center, Kiev, Ukranian SSR, USSR. Another that "borrows" Radio Moscow's transmitters is Radio Vilnius. They have daily programs in English and Lithuanian to North America; the address is Radio Vilnius, Lietuvos Radijas, Konarskio 49, Vilnius, Lithuanian SSR, USSR. Another station broadcasting in English to North America is Radio Yerevan, located in the Armenian SSR. Among the other languages they use are Arabic, Armenian, and French. Their address is Radio Yerevan, 5 Mravian Str, Yerevan, Armenian SSR, USSR. If you're interested in Soviet affairs, it can be interesting to compare the content of the broadcasts from these three stations to that of Radio Moscow.

One other "local" international broadcaster in the USSR can be heard in English. This is Radio Tashkent, located in the Uzbek SSR. It does not transmit to North America, but can be heard in English around 1200–1230 UTC in the 31-meter and 25-meter bands. Their address is Radio Tashkent, Khorezmskaya 49, Tashkent, Uzbekistan SSR, USSR.

There are other international broadcasters in such Soviet republics as the Latvian SSR, Byelorussian SSR, and Kazakhstan SSR. However, these use only the languages of the various republics plus a few languages of nearby nations (such as Arabic, Persian, and Swedish). Most of these also verify listener reports, and many top DXers find hearing such stations and getting QSLs a major challenge.

Regardless of your feelings toward the Soviet Union and its policies, it's obvious that its influence on world events is profound. Listening to Radio Moscow and other international stations from the USSR allows you to hear Soviet positions on various issues first-hand and usually in greater detail than found in much of American media. Whether or not you agree with such views, or even believe them, is up to you and your judgment. Students of Soviet affairs and politics will find the various USSR international broadcasters an endless source of material for study and discussion.

RADIO NEDERLAND

The Netherlands is not a large country in terms of geography, economic power, or military power, but it's a giant when it comes to "information power" through Radio Nederland. Radio Nederland has long been a favorite of SWLs. Its programs are consistently entertaining and informative, and it has shown a special interest in SWLing for years. Moreover, it is a technically innovative broadcaster. Radio Nederland was one of the first stations to experiment with SSB transmission of programs and its transmitter facilities, including relay stations at Madagascar and in the Caribbean, are among the most "state of the art" of any international broadcaster. The Netherlands's commitment to international broadcasting was underscored in 1985, when Radio Nederland placed in operation a massive new transmitter site in the Flevopolder area of the country. The land was reclaimed from the sea, and the facility is actually *below* sea level.

Radio Nederland uses only nine languages, and puts most of its resources into delivering good signals and programs in those languages. All programs begin with news and commentary followed by daily features. Among these will be programs about life in the Netherlands, interviews, pop music, and listener mailbag programs. On Sundays (in UTC), the format is devoted to the "Happy Station Program," currently hosted by Tom Meyer. This is a multilingual program of music and features with the emphasis on light, upbeat entertainment.

Radio Nederland has a special program for SWLs and DXers titled "SW Communications Magazine." This program features reports from well-known and experienced SWLs in various parts of the world; each report is taped by the SWL and includes samples of the stations being heard at his or her location. There are also propagation forecasts, reports on solar activity, and technical features and discussions. Radio Nederland's interest in SWLs does not stop with this program. Radio Nederland offers several pamphlets on various aspects of SWLing to listeners; a complete list is available by requesting a copy of their "Listeners Services Catalogue."

Radio Nederland

Radio Nederland's various relay sites can make it difficult to determine just which location you're tuned to. Fortunately, transmitter sites are indicated in Radio Nederland's program schedule and are announced when each transmitter signs on or regional service (such as to North America) begins. Each transmitter site tends to serve adjacent areas. For example, Bonaire is primarily used for programs to the Americas, while Madagascar is used for broadcasts to Africa and Asia. For the evening broadcasts to North America, most transmissions are from Bonaire, although Radio Nederland has used one frequency from the new Flevopolder site. You won't have much difficulty recognizing the signals from Bonaire; they are quite strong and have little fading, since most of North America is within "single hop" range of Bonaire.

As you might expect, Radio Nederland also likes letters and reports from SWLs. They prefer comments on programming more than data on reception, as they have a network of regular monitors worldwide who contribute such information. However, they do send out colorful QSL cards in response to correct reception reports (although they ask you keep your requests to one QSL card per

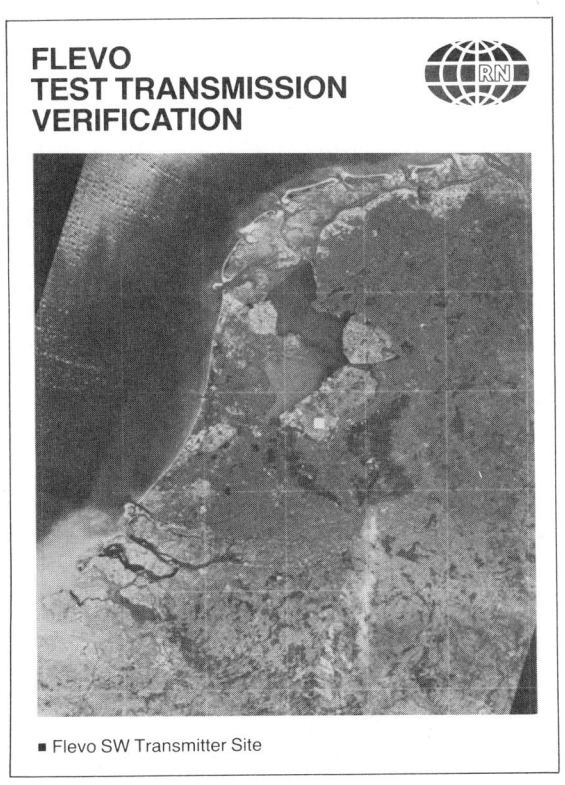

Figure 6.2 QSL from Radio Nederland for test of their new transmitter facilities.

month to help conserve their resources). They even have a leaflet, "Writing Useful Reception Reports," which they gladly send out on request.

Radio Nederland's address is P. O. Box 222, 1200 JG Hilversum, The Netherlands. They have also thoughtfully established several postal drops throughout the world where listeners can send letters without having to pay for overseas postage; these letters are then shipped to Hilversum in bulk on a regular basis. For listeners in the United States and Canada, letters and reports can be sent to P. O. Box 274, Westhill, Ontario, MIE 4R5, Canada. Radio Nederland's English Section has even established a 24-hour automatic answering service for listener comments and questions; you can reach it by dialing the appropriate country code for the Netherlands and 35-18700.

Radio Nederland uses a beautiful interval signal consisting of an old Dutch folk tune played on a carillon and celeste. If you're like many SWLs, that sound—and Radio Nederland—will soon become quite familiar to you.

RADIO FRANCE INTERNATIONALE

Radio France Internationale (RFI) is an example of how what is not said—or transmitted—on SW can often be more revealing than what is. As this was being written, Radio France Internationale was on the air almost six hours each day to North America. Unfortunately, not many listeners in North America could fully enjoy its programs, since they were all in French. That's been the situation for many years as France has resolutely refused to program in English to North America. However, considerable English is used in their transmissions to Africa, and even their transmissions to Latin America, as the time of this writing, included 15 minutes of English daily. In addition to French and English, RFI uses six other languages, including Russian, Polish, and Romanian. Thus, Radio France Internationale gives us the puzzling spectacle of a major world power being upstaged in international broadcasting by much smaller nations and being totally without English service to North America.

Radio France Internationale has long had a particular interest in programming for Africa, a situation not too surprising in view of France's long colonial history on that continent and continuing relationships with many of its former colonies. RFI's programming to Africa is different from most international broadcasting in that it deals more with the target area, Africa, than it does with France. For years, it broadcast an hour-long English program called "Paris Calling Africa" which dealt almost exclusively with African news and events. This was later changed to a format of alternating half-hours of French and English programs. The emphasis is still on African news, however, and RFI's programs to Africa should prove useful to those interested in African affairs.

RFI's English programs to Africa and other areas have a delightful spontaneity to them. Many are apparently done live or in "one take." Thus, you'll hear announcers abruptly interrupt their newscasts for a "hot item" which just arrived.

Radio France Internationale 111

(The importance of most of these "flashes" escapes your author; usually they are trivial.) At other times you can hear announcers fumble for a missing paper or misplaced items; you can sometimes hear announcers telling other persons in the studio, "I can't find it! Where is it?" Compared to the methodical, carefully rehearsed presentations of most other international broadcasters, Radio France Internationale seems warm and very human. In addition to gaffes, RFI also features reviews of the French press (invariably the Paris newspapers) in its English language programs along with French music.

French music can also be enjoyed, even if you don't speak French, during RFI's French broadcasts. Moreover, if you're studying French in school (or want to resurrect what you studied years ago), then RFI is an excellent source of contemporary idiomatic French. (It's surprising that more schools and educators haven't taken advantage of the enormous potential of SW for language study.)

While Radio France Internationale does not go out of its way to solicit letters and reception reports, it does send out colorful QSL cards bearing the logo of its parent agency, Telediffusion de France. Their address is Radio France Internationale, P. O. Box 9516, 75762 Paris Cedex 16, France. As an advertisement for RFI in a recent SW publication said, "Wherever you are—provided you know how to use your receiver—you can hear Radio France Internationale." That

Figure 6.3 Colorful QSL from Radio France Internationale.

line sums up RFI and its attitude in a nutshell. You may find yourself feeling toward it like many Anglophones feel toward the French themselves—one minute you're delighted, the next you're exasperated.

RADIO BEIJING

Perhaps no station has better reflected the changing nature of a country than China's Radio Beijing (formerly Radio Peking). In the late 1960s, at the height of the Great Cultural Revolution, Radio Peking was a shrill, incessant source of revolution and the thought of Mao Tse-tung. Each program opened with a quote from Chairman Mao, and entire programs were devoted to readings from his famous "Little Red Book." When not repeating the latest thoughts from the "Great Helmsman" (as Mao was known in China), Radio Peking treated SWLs to performances of revolutionary operas such as *The Red Detachment of Women*.

As China began to open itself to the West, Radio Peking's line began to moderate. With the establishment of diplomatic relations with the United States and the downfall of the "Gang of Four," Radio Peking/Beijing became one of the most interesting of all international broadcasters. ("Beijing" reflects a new system of spelling with the Roman alphabet now adopted in China.)

Radio Beijing is far from an independent broadcaster, however. It is still very much an instrument of the Chinese government, and its purpose is the same as Radio Moscow's—to further national policy and objectives overseas. Nor will you find Radio Beijing's news and commentaries much more honest and candid than those on Radio Moscow (particularly where internal Chinese events are concerned). But where Radio Moscow is often offensive or just plain dull, Radio Beijing is usually lively and interesting, and it leaves you with positive feelings toward China and its people. While Radio Moscow may have more frequencies for and stronger signals in North America, there's little question that Radio Beijing is more popular among listeners.

Radio Beijing uses the first 19 notes from "The East Is Red" for its interval signal, and intersperses it with identification announcements in the languages of the target area. Each program opens with a full instrumental version of "The East Is Red" followed by news and a commentary of the news. While world events are covered, there is also a significant number of news items pertaining to China itself. This can be a rich resource of material on Chinese affairs and politics, and can give clues as to the direction of current Chinese policy on a variety of domestic and international issues.

Following the opening news are some of the interesting new features that Radio Beijing has added in recent years. One continuing series is "China in Construction," which has used other names in the past (such as "China Reconstructs," which was also the title of a magazine distributed to listeners). Actually, "construction" in China today means more than building new dams, housing, airports, and factories; it refers to all of China's modernization efforts. In the late 1960s, programs on construction were little more than testimonials on how ardent study and application of the thoughts of Chairman Mao enabled construction brigades to complete works in record time and with fewer materials. Today, you'll find a much more honest discussion of the difficulties involved in attempting to adapt Western technology and management methods to China. Several of China's new economic experiments, such as a limited free market, have been explored on "China in Construction." There have been frank admissions of some of the difficulties—corruption, black marketeering, and the like—involved. This is in marked contrast to Radio Moscow's position that almost nothing ever goes wrong in the USSR, and certainly nothing that might be a product of the system itself.

Radio Beijing has tapped into China's long and rich cultural traditions for programming material. There have been programs based on Chinese folk tales, archeology, and history (covering such as the origin of paper in China). Both contemporary and traditional Chinese music is featured, along with post-Cultural Revolution drama and comedy (in English translation). One regular segment was by an American woman living in Beijing, and gave her perspectives and impressions on contemporary China.

Unfortunately, Radio Beijing is not always easy to receive in North America, since most signal paths must cross the North Polar regions. Radio Beijing also lacks the relay bases used by other broadcasters; the only relay they ever used was via Albania's Radio Tirana. This relay was abruptly terminated when China and Albania broke diplomatic relations in the late 1970s. (Faithful to the end, Radio Tirana even relayed then-Radio Peking's announcement of the break in relations and scathing denunciations of Albania.) The evening North American service primarily utilizes the 19-meter and 16-meter bands, with some use of the 25-meter and 31-meter bands during the winter. Reception during the evening in eastern North America can be erratic, depending on the state of the Earth's geomagnetic field. Better reception is often possible around 1000–1500 UTC in the 31-meter and 49-meter bands. Like other international broadcasters, Radio Beijing switches frequencies to compensate for seasonal propagational changes. However, some frequencies, such as 9860 and 9820 kHz, have been used for years. Radio Beijing is also famous for using some oddball channels; 4130, 4200, 6520, 6860, 6995, 10865, and 11455 kHz are among those used for English transmissions. Try for these channels around your local sunrise and sunset.

Radio Beijing welcomes letters and reception reports from SWLs. In addition to colorful QSL cards, you can expect to receive magazines, booklets, calendars, and other items if you're a regular reporter to them. Their address is simply Radio Beijing, Fu Xin Men, Beijing, People's Republic of China. Use the full name of the country—Chinese postal authorities have been known to return letters simply addressed to "China."

If you share the usual Western fascination with China, you'll become a fan of Radio Beijing. And, as China continues to evolve into a superpower, more surprises can be expected at Radio Beijing.

RADIO TIRANA

And what of Radio Peking's former relay partner, Radio Tirana? Suffice it to say that Radio Tirana may well be the most bizarre of all international broadcasters. But what else can you expect from a nation like Albania, which shuns alliances—or even diplomatic relations—with the West, the Soviet bloc, China, and most of the Third World and considers Stalin as the last Soviet leader that wasn't too "soft?"

Radio Tirana's nonstop yawping sounds like a bad skit from the old "Saturday Night Live" television program. You can catch newscasts in which American imperialists and Soviet revisionists are blasted in the same sentence. You can hear features on the glories of Albanian communism, such as the opening of Albania's first light-bulb factory. And you can catch the latest details on economic development goals set by the Albanian Communist Party Congress. The overall impression is of an extremely xenophobic, ethnocentric nation—which Albania is—whose international broadcasting station couldn't care less about what their listeners might be interested in.

Radio Tirana did receive some powerful transmitters from China before their eventual break, so Radio Tirana manages to put strong signals into North America. However, engineering practices are not very high at Radio Tirana; their frequencies tend to drift slowly, and the audio quality is often poor. One can only surmise that Radio Tirana expects SWLs to put up with such difficulties to catch the pearls of wisdom it dispenses.

Radio Tirana's frequency selections are sometimes contemptuous of international agreements. For example, it uses 7030, 7065, 7075, 7080, and 7090 kHz; the 7000–7100-kHz range is supposedly reserved for amateur radio exclusively worldwide. Two strange frequencies used are 14320 and 16230 kHz. For North American listeners, 6200 kHz is often used in the evenings for English and Albanian programs.

One reason you might want to listen is to get a QSL card. Radio Tirana verifies reception reports on an irregular basis; the QSL may come quite late (over a year) after the report was mailed, or it may take a second or third report to nail one of their cards. You can send your reports to Rruga Ismail Qemali, Tirana, Albania. Enjoy!

EVANGELICAL BROADCASTING: HCJB AND TRANS WORLD RADIO

As mentioned previously, various Christian evangelical groups make extensive use of SW broadcasting. This is in marked contrast to major religious groups; for example, despite the massive wealth accumulated in recent years by various Islamic countries, there has been no effort to establish stations and broadcasting organizations to further the spread of Islam. The effectiveness of evangelical broadcasting is hard to measure and is the subject of wide debate. Regardless of its actual impact, evangelical broadcasting has steadily grown over the past two decades, and more religious organizations seem ready to devote time and resources to broadcasting. Such broadcasters will likely closely follow the two "prototypes" of international broadcasting: station HCJB, "the Voice of the Andes," located in Quito, Ecuador, and Trans World Radio (TWR), a New Jersey organization which operates stations from various locations such as Monaco and Guam.

HCJB is operated by World Radio Missionary Fellowship, a group headquartered in Opa Locka, Florida. While the call letters are drawn from the block allocated to Ecuador, the selection was not accidental; the call sign stands for *h*eralding *C*hrist *J*esus's *b*lessings. The scope of HCJB's operations is greater than some government-financed broadcasters. HCJB broadcasts in ten languages, including Japanese, Russian, and Swedish. Transmitting facilities include a 500-KW unit, and HCJB has even constructed its own hydroelectric station to provide power for its facilities. It also transmits in Quechua, a language used by Indians living in the Andes, and maintains a station on 690 kHz which programs exclusively in Spanish for a local audience.

HCJB has also made a strenuous effort to maintain good relations with SWLs. It has a weekly program specifically for SWLs and DXers called "DX Party Line" which is produced in English, German, Swedish, Japanese, and French. It also sponsors its own listeners' club known as ANDEX (for "Andes DXers"). HCJB welcomes listener mail and reception reports, and issues a series of colorful QSL cards, which are changed several times per year.

If you want to learn about the politics and culture of Ecuador and South America, you'll be disappointed by HCJB. Politics is strictly eschewed by HCJB, but this is to be expected of a private broadcaster. Content relating to Ecuador or Latin America is subordinate to the prime mission of evangelism. Thus, you'll find many programs, such as "Guide for Family Living," which are similar to those found on local religious broadcasters on AM and FM. Even those programs which are ostensibly not religious, such as "DX Party Line," manage to work in some moments connected with evangelical Christianity. The approach is generally "soft sell," but the prime purpose behind HCJB is to convert listeners to the religious viewpoints of the station's sponsors. There is nothing sinister in this, of course, but it does explain why its program content is sometimes indifferent to Ecuador and South America.

HCJB's Spanish service, particularly via the 690-kHz outlet, does have a strong "service orientation" for its audience. There is a greater proportion of educational and cultural content.

HCJB welcomes reports to Casilla 691, Quito, Ecuador. In addition to a QSL card, SWLs can expect to receive a copy of HCJB's latest program schedule and a religious tract or two in response to a correct report. While transmission schedules and frequencies do change seasonally, the powerful signals HCJB delivers into North America make it an easy catch.

Trans World Radio is exclusively an evangelical broadcaster, using several different transmitter sites worldwide. One is located in Monaco, a tiny nation wedged between France and Italy, and transmits religious programs in a variety of European languages, including Croatian, Hungarian, Latvian, Serbian, and White Russian. Shortwave transmitter powers of up to 500 KW are used, but all are intended for European reception. However, it has English programs for Europe between approximately 0500–1100 UTC in the 41-meter, 31-meter, and 25-meter bands, and these can often be received in North America. You won't hear any news or information about Monaco on TWR; all programs are strictly religious. (This is the case with all other TWR transmitter sites.) In fact, unless you are interested in religious programming to begin with, there will probably be little reason for you to become a regular listener to any TWR outlet.

The prime reason most SWLs tune to TWR at Monaco is to secure a QSL card for that particular country. In English, you'll note that the station identifies itself as transmitting from Monte Carlo rather than Monaco; Monte Carlo is Monaco's capital (for all practical purposes, Monte Carlo *is* Monaco) and home to the casinos that draw tourists and produce wealth for the tiny principality. You can send your reception reports to Trans World Radio, B. P. 349, Monte Carlo-98007, Monaco.

Incidentally, the question of whether you can really hear Monaco via TWR is disputed by some SWLs, since the transmitters for TWR are actually located on French soil. A tiny, but vocal, minority in some SWL clubs along with the publisher of one commercial SWLing publication maintain that all TWR/Monaco broadcasts should be counted as from France rather than Monaco itself. Most SWLs, however, continue to use the location where a station is licensed or authorized as the criterion for determining location unless the geographic separation between authorized location and transmitter site is significant. In the case of Monaco, the separation is minor.

The TWR outlet you're most likely to hear is the one located on the island of Bonaire in the Netherlands Antilles, a group of islands off the coast of Venezuela. In addition to SW, TWR also transmits on the AM broadcast band, using a 500-KW transmitter on 800 kHz. English is often scheduled from 0230–0500 UTC, and you may be able to hear TWR on 800 kHz if QRM permits. (Listeners in the central and western parts of the United States may encounter interference from Canadian and Mexican stations on this channel.) As you might expect, TWR at Bonaire broadcasts in the languages spoken in the Americas, such as Spanish, English, and Portuguese, but it also includes German and Russian as well. English transmissions to North America on SW are usually scheduled around 0400–0500 UTC and again at 1100–1245 UTC; usually a frequency in the 25-meter band is used. All transmissions are clearly identified as coming from Bonaire, and it does accept QSL correct reception reports sent to Trans World Radio, Bonaire, Netherlands Antilles. You may recognize the station by its interval signal consisting of the opening notes of "Stand Up, Stand Up for Jesus" played instrumentally.

Another TWR outlet operates from the U.S. territory of Guam. Since this station operates from the jurisdiction of the United States, it has call letters—KTWR. The SW service is intended for reception in Asia, and makes extensive use of Asian languages such as Bengali, Cantonese, Indonesian, Japanese, and Tamil. There are also English programs as well.

Unlike some other TWR outlets, KTWR does have features of interest to the SWL and DXer. At the time this was written, it carried "DX Listener's Log" in both English and Japanese, along with a weekly mailbag program. Reception reports, along with requests for program schedules, go to KTWR, P. O. Box CC, Agana, Guam, 96910. Their interval signal is the first few notes of "We've a Story to Tell to the Nations" played on an organ.

A more difficult TWR outlet to hear in North America is their station located in Swaziland. This site is used for programming to Africa, and its times of operation are not ideally suited for North American reception. Moreover, the most powerful transmitter used in Swaziland is only 100 KW and more typically only 25 KW is used; several frequencies in the 90-meter and 60-meter tropical broadcasting bands are employed. Besides English and French, you can hear such exotic languages as Chewa, Lingala, Ndebele, Sotho, Tshwa, and Zulu over TWR in Swaziland. You may find it easier to identify the station by its interval signal, which is the same as that used for KTWR but played with bells. The address for reception reports and program schedules is P. O. Box 64, Manzini, Swaziland.

TWR also maintains another transmitter location in Sri Lanka (formerly known as Ceylon) and also broadcasts over Radio Monte Carlo's relay station on the island of Cyprus. However, both of these operations exclusively take place on the AM broadcast band, and reception in North America is unlikely.

As mentioned at the beginning of this section, religious broadcasting activity has been increasing rapidly in recent years. Regardless of one's opinions about such efforts or programs, it has been the force behind the establishment of several new international broadcasting stations in the United States, and shows no signs of lessening its impact on SW radio in the foreseeable future.

VATICAN RADIO

One might think that Vatican Radio would be in the same category as HCJB and TWR. However, HCJB and TWR broadcast to convert listeners while Vatican Radio, in effect, broadcasts to the already converted. Vatican Radio's programs are usually brief—a half hour or less—and deal primarily with Church affairs and news. Programs are presented in over 30 languages, including Esperanto. Other broadcasts include the celebration of the Mass in various languages, including Polish and Latin.

Vatican Radio's broadcasts to North America open with "Christus Vincit" played on a celeste followed by the phrase "Laudetur Jesus Christus. Praised be Jesus Christ. This is Vatican Radio." This is usually followed by news and commentary, and then a feature dealing with some aspect of the Church or Vatican City. (Interviews with American visitors to Vatican City are a frequent feature.) At other times, Vatican Radio broadcasts live addresses by the Pope. The normal schedule of Vatican Radio goes out the window when the Papacy is vacant; the first word that a new Pope has been elected often comes from Vatican Radio. (In fact, published reports have stated that Vatican Radio had a "spy" with a concealed transmitter in the conclaves that elected Popes John Paul I and John Paul II to ensure that they would be the first to announce the new Pope.)

There is usually one daily transmission for North America beginning at 0050 UTC; one frequency each in the 49-meter, 31-meter, and 25-meter bands is used. Transmitter powers range up to 500 KW, and signal strength is usually excellent.

Vatican Radio welcomes listener reports and responds with QSL cards that are changed on a regular basis. They also send out a multilingual program schedule to reporters. Their address is simply Vatican Radio, Vatican City.

The status of Vatican City is questioned by some of the same SWLs who

Figure 6.4 Vatican Radio is perhaps the only SW broadcaster who can show almost its entire nation on a picture QSL!

question the status of TWR in Monaco. Vatican City itself is obviously too small to contain a major broadcasting facility; in fact, the actual transmitting site is located some distance from Rome. However, under agreement between Vatican City and Italy, the transmitter site has extraterritorial status from Italy and is administered by Vatican City. A few SWLs disregard this and insist that all Vatican City transmissions originate from Italy. The majority of SWLs, however, consider the agreement valid and consider Vatican City's transmitter location separate from Italy.

RADIO CANADA INTERNATIONAL

Canada is a nation in an unusual situation. It shares the world's longest undefended border with the United States, and is the United States's largest trading partner. Yet awareness of Canadian issues and interests in most of the United States is quite low. Thus, it's hardly surprising that more hours of programming are directed to the United States by Radio Canada International than to any other area.

Radio Canada International (RCI) also seeks to articulate Canadian issues, culture, and concerns in other parts of the world. It currently uses eleven languages for its programs, including Hungarian, Japanese, Russian, and Ukrainian. The prime transmitting facility is located at Sackville, New Brunswick, and uses transmitters of up to 250 KW. RCI's programs are relayed in Europe by Great Britain's BBC and Portugal's Radio Trans Europe, a commercial SW station that rents its facilities to various broadcasters and organizations (including Radio Japan). RCI programs are also broadcast over Radio Tanpa, a domestic Japanese SW broadcaster, and by Radio Antilles, an AM station located on the Caribbean island of Montserrat.

Radio Canada International seems to understand the United States SWL community better than any other broadcaster. Partly this is because Canadian and American SWLs are, like the United States and Canada themselves, very much alike (although important differences do exist, of course). An even bigger reason may be RCI's efforts to expand its links to the SWL community. Its popular weekly program for SWLs, "Shortwave Listeners' Digest," contains the latest SW news and special features (such as a course on how to recognize various foreign languages and reviews of new SW receivers). RCI personnel, such as Ian McFarland, frequently contribute to SWL clubs and publications and often attend SWL club conventions. In fact, RCI has hosted an Association of North American Radio Clubs (ANARC) convention at RCI's headquarters in Montreal.

Radio Canada International has even come up with an innovative solution to the problem of answering reception reports. Like many broadcasters, RCI was faced with increasing costs involved with processing requests for QSL cards. Some stations have begun to send out nonspecific acknowledgment cards, rather than true QSL cards, in reply to reports. RCI instead sends out a blank QSL card with their summer/autumn program schedule to all SWLs on their mailing list. Listeners desiring a QSL return it to RCI with a report. RCI does check all reports and, if the reports are correct, returns the validated QSL. This approach allows RCI to better utilize its resources and still gives SWLs an acceptable QSL card. To get on their program schedule mailing list, write to RCI, P. O. Box 6000, Montreal, Quebec, H3C 3A8, Canada. Of course, RCI appreciates reports and comments on their programs even if you're not seeking a QSL.

RCI's programs are among the best produced and most consistently interesting you'll find on SW. One is "Canada a la Carte," which focuses on a variety of Canadian topics; they range from health care to high-tech industries to regional cuisine of the Canadian provinces. News programs include "The World at Six" (which, despite the title, may be aired at such times as 0000 an 0200 UTC) as well as newscasts on the hour when "The World at Six" is not presented. RCI also uses some programs from the domestic Canadian Broadcasting Corporation (CBC) such as "Sunday Morning." As you might expect, it is broadcast only on Sunday mornings and is primarily intended for Canadians abroad. It is currently three hours long, and American listeners will find it an excellent guide to issues and events of concern to Canadians. RCI also relays CBC programs in English, French, and Eskimo languages to northern Quebec.

WRNO, KYOI, AND AMERICAN COMMERCIAL SHORTWAVE BROADCASTING

Domestic shortwave broadcasting has never existed to any real extent in the United States, mainly because there are few areas (with the exception of northern Alaska) where several AM and FM stations cannot be received easily. Moreover, the difficulty of using and tuning the older models of SW receivers helps ensure that most people are content with whatever they can find on their local stations.

Commercial international broadcasters have not fared much better. The amount of money it takes to construct and operate a SW broadcasting facility is considerable; investment in a local AM, FM, or TV station is far less risky. Moreover, the ad revenues necessary to sustain such a station are considerable, and the benefits of advertising on such a station would be difficult to "sell" to prospective ad buyers. As a result, commercial SW broadcasting was virtually unknown after World War II. (The only exception was a curious operation known as Radio New York Worldwide, whose heyday was the 1960s. This station will be discussed later, but it seems that substantial portions of its operating expenses were provided covertly by the U.S. government.) Private international broadcasting from the U.S. was the province of stations such as KGEI, San Francisco, and WINB, in Pennsylvania, whose content was (and still is) strictly evangelical Christian programming.

By the late 1970s, the situation had started to change. The growing affluence of the rest of the world meant that advertising possibilities were greater;

a long-term investment in SW broadcasting facilities was not totally out of the question, especially if some other forms of revenue (such as selling air time) could be generated. The revolution of SW receiver technology had spread far enough so that finding a specific station or frequency was no longer a chore. The convergence of these factors resulted in some broadcasting entrepreneurs seriously examining the possibility of commercial SW broadcasting.

The first commercial station to accept the challenge was WRNO in New Orleans, Louisiana. WRNO was already a successful FM station with a format of contemporary music. Its plan for international broadcasting was simple but effective. It would sell some air time each day to religious organizations to provide a "floor" of revenue for the SW station and then "simulcast" the FM station the rest of the day. For SWLs, it added a DX program, "World of Radio," hosted by well-known SWL Glenn Hauser. For more casual SWLs, WRNO also began a worldwide call-in request program. WRNO also lets European SWLs follow the New Orleans Saints football team via its live broadcasts of their games (not a bad idea, when one considers the amazing growth in the popularity of American football in Britain and France).

At the present time, WRNO operates from 1600–0600 UTC daily. On weekdays, the first two hours (1600–1800 UTC) are devoted to paid religious programs and the remainder of the day is a relay of the FM station. On Saturday and Sunday, WRNO broadcasts paid religious programs, public service programs, and features such as "World of Radio." Frequencies used included 6185, 7355, 11965, and 15420 kHz, although these may have changed by the time you read this. Reception reports and program schedules are available from WRNO, P. O. Box 100, New Orleans, Louisiana, 70181.

The next commercial SW effort was vastly different from WRNO. KYOI is located on the Pacific island of Saipan in the U.S. territory of the Mariana Islands. The premise behind KYOI was twofold: (1) Japan has perhaps the greatest level of SWLing interest in the world, and (2) Japan's youth are intensely interested in Western rock music. Thus, KYOI broadcasts nothing but rock music—with Japanese announcers—24 hours each day to Japan. Since KYOI optimizes frequencies, transmission times, and antenna direction of best reception in Japan, it is not an easy catch in North America. But if you do hear them, or would like their latest program schedule, you can write them at KYOI, P. O. Box 1387, Saipan, CM, 96950.

WRNO and KYOI have sparked a variety of plans for other new SW broadcasting stations. Perhaps the most ambitious was for "NDXE Worldwide," to be located near Opelika, Alabama. Their plans called for 24-hour broadcasting consisting of news, sports, music, and (to quote their literature) "in-depth coverage of financial, scientific, and technological developments." Besides English, such languages as French, Spanish, Portuguese, and Japanese were to be used. Plans called for transmitters of up to 500 KW, capable of AM stereo operation, to be used. NDXE even had a survey card bound into the 1986 edition of the *World Radio TV Handbook* to ascertain SWL desires in programming and station operations. Similar plans have been announced, and some new religious SW broadcasters, such as KCBI in Dallas, Texas, have started operations.

While it's unlikely that the United States will ever have private SW broadcasting on the scale it is found in Latin American nations, it's clear that more new stations will be heard in the years ahead.

RADIO RSA

Radio RSA bills itself as "The Voice of South Africa," and is the overseas broadcasting arm of the South African Broadcasting Corporation (SABC), a government agency. SABC is responsible for all broadcasting in the Republic of South Africa; there are no privately owned stations. In addition to Radio RSA, SABC also operates extensive domestic networks on AM, FM, and SW.

Radio RSA, as an operation of the South African government, naturally tries to present South Africa in the most favorable light possible and is a staunch defender of the nation's policies, including apartheid. The defense of apartheid includes more than just program content. For example, a recent program schedule from Radio RSA included a group photograph of the staff employed by the English language section. The photograph included four black staff members, giving the clear impression that Radio RSA itself does not discriminate in its hiring or employment practices. Another schedule showed a black student receiving a university degree; omitted was the information that the university in question was all-black and the student could not have attended a university for whites had he so desired.

Radio RSA makes a strong effort to cultivate goodwill among the international SWL community. Reception reports are promptly answered with attractive QSLs, and designs are changed frequently. Station pennants are also freely sent. Radio RSA also broadcasts its own DX program, "DX Corner," in English and five other languages. One innovation Radio RSA began several years ago was an international telephone call-in program on New Year's Eve. As you might expect, most callers are simply persons who want to hear their own voices over a major SW broadcaster. However, Radio RSA has previously tried to subtly interpret the results of the show as expressions of support for Radio RSA and the South African government. Other measures of listener response, such as letters received each year, have been interpreted in similar fashion.

Radio RSA uses eleven languages for its external services along with transmitters rated at up to 500 KW. Its transmissions to North America in English usually begin around 0200 UTC and continue until approximately 0300; commonly used frequencies have included 5980 and 9615 kHz. A current schedule can be obtained from Radio RSA, P. O. Box 4559, Johannesburg 2000, South Africa. Reception reports can also be sent to this address.

Radio RSA's interval signal is one of the most beautiful you'll hear on SW. It is the chirping of the Bokmakierie bird along with the first few bars of the

Figure 6.5 Radio RSA verifies reports with QSLs showing various South African scenes.

South Africa folk song, "Ver in die Wereld Kitty," on acoustic guitar. Each program opens with news. Other features include "Peoples of South Africa," "Letter from South Africa," "Africa Today," and "Touring RSA." Listener letters are answered in the program "P. O. Box 4559." While the approach is not heavy-handed, the programs are a consistent defense of the policies of South Africa.

Like Radio Moscow, one does not have to agree with Radio RSA or the policies of South Africa to recognize the importance of South Africa in world affairs. Students of politics and African affairs will find Radio RSA essential listening in the years ahead.

DEUTSCHE WELLE

Deutsche Welle (DW) is West Germany's international broadcasting service. It is, by any standards, an impressive operation. Currently, 34 languages are used by DW, including such exotic tongues as Dari, Maghreb, Pushto, and Urdu. Its main transmitting facility is at Wertachtal, West Germany, which uses transmitters rated at 500 KW. Deutsche Welle also uses an extensive relay station network. One relay is via Radio Trans Europe in Sines, Portugal (the same relay used by Radio Canada International) and is used for transmissions to the USSR and eastern Europe. This is because jamming transmitters in the target countries lie in the skip zone for some transmissions from Sines. Other DW relays are in the African nation of Rwanda, the island of Malta, on Sri Lanka, and on the Caribbean islands of Montserrat and Antigua. The Antigua facilities are shared with the BBC and are heavily used, along with Montserrat, for transmissions to North America. Many of the other relay sites can be heard in North America as well.

In recent years, Deutsche Welle has made an effort to become the "voice of Europe" as much as the "voice of Germany." Each program opens with news and is followed by "Microphone on Europe," a program with recorded excerpts from persons on various European topics. This is usually followed by features on such topics as classical music, opera, West German life, and the European economy. Deutsche Welle also devotes attention to non-European topics; one recent program was on American millionaires. There is also a monthly DX program in German, English, Japanese, and Spanish.

Each DW broadcast to North America lasts approximately 50 minutes. Their distinctive interval signal is an excerpt from Beethoven's "Fidelio" played on a celeste. Unfortunately, there is no way to tell which Deutsche Welle transmitter site you are listening to without referring to their transmission schedule.

Deutsche Welle does not send out specific QSL cards any more; instead, a nonspecific acknowledgment of a report is sent out. (The only exceptions are certain anniversaries or other special events.) DW also requests that listeners desiring a QSL send a preaddressed, unstamped envelope with reception reports. Reception reports and requests for program schedules go to Deutsche Welle, Postfach 100444, D-5000 Koln 1, West Germany.

Fortunately Deutsche Welle is worth listening to for itself rather than just to get a QSL card. It is emerging as a rival to the BBC as Europe's main shortwave broadcaster. Even those with just a casual interest in what's going on in Europe's major economic power will find DW worth a listen.

VOICE OF AMERICA

It might seem that the Voice of America (VOA) wouldn't be of much interest to American SWLs. Such isn't the case. It is always interesting to see how one's tax revenues are being spent and how the United States tries to present itself abroad.

The Voice of America is the broadcast division of the United States Information Agency (USIA); USIA also distributes print, audio, and video materials about the United States throughout the world. The USIA is headed by a director appointed by the President of the United States; this has resulted in complaints by some VOA staff members and other observers that such appointees have tried to influence the editorial processes of the VOA and other USIA functions to reflect favorably upon the administration and political party currently in power.

The VOA is a massive broadcasting organization, currently producing programs in 42 languages. In the United States itself, transmitter sites at Greenville, North Carolina; Bethany, Ohio; Delano, California; and Dixon, California are used. The Greenville, North Carolina facility is the main VOA transmitting site in the United States, and transmitter powers of up to 500 KW are used from there. The other three domestic sites have transmitters rated at 250 KW. There is another VOA station, at Marathon, Florida, which operates only on the AM broadcast band for programs to Cuba.

The Voice of America also uses an extensive network of relay stations. SW transmitting facilities are maintained by the VOA in West Germany, Greece, Liberia, Morocco, the Philippines, and Sri Lanka. The VOA is also relayed in Great Britain by the BBC and in Brazil by Radio Bras, the Brazilian government's SW broadcasting service. In addition, the VOA maintains AM broadcast band relay stations in Antigua and Botswana.

Identifying the various transmitter sites is easy if you tune to the start of a VOA transmission; the sign-on announcement includes the transmitter location being used. The VOA uses several different instrumental versions of "Yankee Doodle" for an interval signal.

Many SWLs find the Voice of America's English programs somewhat dull when compared to some other international broadcasters. (Decide for yourself. Spend an hour listening to Radio Canada International and then an hour listening to the Voice of America.) Part of this arises from the desire of VOA staff to be impartial and objective in both the selection and treatment of topics broadcast. However, the results sound very much like "programming by committee." There is little that could offend or annoy but, unfortunately, often little to catch one's interest.

The VOA's newscasts do enjoy a high reputation for accuracy and honesty. It also has news and events programs covering the target areas it broadcasts to, such as "Caribbean Report" and "Report to the Americas." The VOA also broadcasts considerable American music—including America's only original music form, jazz—on programs such as "Music Now" and "Music USA." The "Magazine Show" covers a miscellany of American subjects daily.

The Voice of America presents some newscasts and programs in what is known as "special English." Such programs use a limited vocabulary of common English words and verb tenses; the announcers speak very slowly and enunciate carefully. This allows persons with a limited command of English to both understand VOA newscasts and to also improve their English skills.

The VOA does not neglect SWLs. Each Thursday's "Magazine Show" features a segment on SWL subjects. Moreover, the VOA is an excellent verifier of reception reports. Voice of America reception is not always routine. For example, the Sri Lanka relay uses transmitters of only 35 KW, and reception in North America can be a real challenge even for experienced SWLs. The VOA's West German and British relays are the easiest ways to get "full data" QSLs for those two countries, and the VOA is the easiest way to hear and verify such countries as Morocco, Liberia, and the Philippines. Reception reports and requests for program schedules go to Voice of America, U. S. Information Agency, Washington, DC, 20547.

For most American SWLs, SWLing lets us know how other countries view us. The Voice of America lets us see how we view ourselves—at least "officially." (And, unlike other SW broadcasters, if you don't like what you hear on the VOA, you can always write your congressional representatives!)

OTHER INTERNATIONAL BROADCASTERS

This chapter has been little more than a sample of the variety of international broadcasting stations you can hear on shortwave. The latest frequencies and schedules can be found in the latest edition of the *World Radio TV Handbook*, in SWL club bulletins and commercial publications, or by writing the various major stations whose addresses can be found in Table 6-1.

As stated before, some SWLs hone their listening skills on major international broadcasters and then move on to more challenging DX targets. Other SWLs confine their listening to their favorite international broadcasters. Most SWLs get pleasure out of both worlds, tuning for rare DX stations some of the time but returning to their favorite stations and programs on a regular basis. The continuing growth of international broadcasting will ensure that there will be plenty for everyone to listen to.

TABLE 6-1 ADDRESSES OF MAJOR INTERNATIONAL BROADCASTERS

ARGENTINA:	RAE English Services, Casilla 555, 1000 Buenos Aires
AUSTRALIA:	Radio Australia, GPO Box 428G, Melbourne 3001
AUSTRIA:	Radio Austria International, A-1136 Vienna
BELGIUM:	RTBF International Services, Box 202, B-1040 Brussels
BULGARIA:	Radio Sofia, 1421 Sofia 21
CUBA:	Radio Havana, Apartado 7026, Havana
CZECHOSLOVAKIA:	Radio Prague, Praha 2, 12099 Vinohradska 12
EGYPT:	Egyptian Radio and TV, P.O. Box 1186, Cairo
FINLAND:	Radio Finland, Box 10, SF-00241 Helsinki 24
GERMANY (EAST):	Radio Berlin International, 1160 Berlin
HUNGARY:	Radio Budapest, Brody Sandor 5-7, H-1800 Budapest
ISRAEL:	Kol Israel, Overseas Services, P.O. Box 1082, Jerusalem
ITALY:	RAI English Service, Viale Mazzini 14,00195 Roma
JAPAN:	Radio Japan, NHK, 2-2-1 Jinnan, Shibuya-ku, Tokyo
NORWAY:	Radio Norway International, N-Oslo 3
POLAND:	Radio Polonia, PL 00-950, Warszawa
PORTUGAL:	Radio Portugal, Av. Eng. Duarte Pacheco 5,1000 Lisboa
ROMANIA:	Radio Bucharest, P.O. Box 1-111, Bucharest
SPAIN:	Spanish Foreign Radio, Apartado 156.202, 28080 Madrid
SWEDEN:	Radio Sweden International, S-10510 Stockholm
SWITZERLAND:	Swiss Radio International, Giacomettistrasse 1, CH-3000 Berne
TURKEY:	Voice of Turkey, P.O. Box 333, Yenisehir, Ankara

7

DOMESTIC SHORTWAVE BROADCASTING

Domestic SW broadcasting is in stark contrast to the powerful signals and carefully polished programs of international broadcasting. Intended for reception within the country where the stations are located, domestic SW broadcasters are much more challenging, and often difficult, listening fare than international broadcasters. Yet many SWLs find them more rewarding in terms of programming, and DXers find the ultimate tests of their skill among them.

Your first efforts to tune domestic SW stations may discourage you somewhat. You won't hear the loud, thumping signals you expect from major international broadcasters. Many international stations pump out 500 KW of transmitter power. By contrast, a 50-KW domestic SW station is a powerhouse, and transmitter powers of 5 to 10 KW are far more common. Even lower powers—even as little as 250 W—are often found. Moreover, most domestic stations operate on lower frequencies, particularly the 120-meter, 90-meter, and 60-meter tropical broadcasting bands. Such bands are often well below the MUF for a given path and suffer more from absorption from the D-layer and E-layer of the ionosphere. Domestic stations operate according to the needs of their local population, not when conditions are best for reception thousands of miles away. All these factors mean that even the most die-hard SWL is forced to become something of a DXer when tuning domestic SW stations.

The language barrier can also be a problem. You'll hear very little English on domestic SW stations. Instead, expect to hear a lot of Spanish, French, and Portuguese. But you'll also hear Indonesian, Tahitian, Lingala, Swahili, Hindustani, Malay, and other languages used by the populations where the stations are located. Trying to identify a given station can be difficult, especially in those cases where several stations broadcasting in the same language are found on the same frequency. But if you spend much time listening to domestic SW stations, you'll gradually develop the ability to recognize (if not understand) several languages. And eventually you'll learn the equivalent phrase for "This is radio. . ." in many languages. You may actually pick up several different phrases and expressions in some of the more common languages. Of course, if you're studying a foreign language, or want to keep your knowledge of one fresh, it's hard to beat the practice material afforded by domestic SW stations.

The rewards of listening to domestic stations extend beyond increased listening skill and language awareness. The music you can hear is alone worth the listening effort. Ever heard what Chinese opera sounds like or the so-called African "high life" type of music? How about authentic South Seas music played with "slide" guitars? Those are just some of the music types you can hear on domestic SW broadcasters. You'll discover that music can be a valuable clue as to which nation the station you're listening to is located in. For example, you'll come to associate "ranchero" music (which sounds much like traditional American country and western music) with Mexico, while "merengue" will indicate you're probably listening to a station in the Dominican Republic. Even nonmusical programs can be enjoyed. You don't have to understand the language to get a kick out of hearing a soccer match from a Latin American station or a radio soap opera in a foreign language.

Domestic stations change frequencies and hours of operation much less often than international SW broadcasters. In fact, many stations have operated on the same frequencies for decades. Seasonal frequency changes common among international broadcasters are almost unknown among domestic broadcasters. When a change is made in a station's frequency, it tends to be relatively permanent (that is, lasting several years). Some domestic SW stations accidentally change frequencies when their transmitters "drift" off their assigned frequencies. Such conditions can last only a few days or may be a constant fact of life for a few stations. Of course, new domestic SW stations come on or leave the air each year.

Domestic SW broadcasters may be government-operated, religious, or commercial. Many nations in Asia and Africa, especially those located in the tropics, use shortwave for services to their local population. Not all such nations are underdeveloped; Australia uses SW to reach persons in that nation's often isolated interior. The USSR also finds SW an effective method for broadcasting to isolated communities in that sprawling nation. Even the Canadian Broadcasting Corp. (CBC) operates two low-power SW outlets on the Atlantic and Pacific coasts.

Religious broadcasters also use SW for domestic broadcasting. Some are operated by evangelical Protestants, while others (particularly in Latin America) are operated by the Catholic church. Domestic religious broadcasters often have a strong service orientation toward their listeners, carrying educational programs on such topics as health care for an audience that is often largely illiterate.

Commercial domestic SW broadcasters are plentiful, especially in Latin

America. Such stations are supported by advertising, and you'll hear the names of familiar products (such as Ford and Sony) in foreign-language ads along with some distinctly local products (such as "Jungle Oats" in South Africa and "Inca Cola" in Peru). Commercial domestic SW stations are also found in such nations as Japan and Canada; the latter stations often give SWLs in the rest of the world an idea of what commercial Canadian and American radio sounds like.

The QSL collector faces a more formidable challenge when it comes to extracting a verifying card or letter from many domestic SW stations, especially those in less developed areas of the world. Most domestic SW stations couldn't care less how well they are heard in a distant country; if their local listeners can hear them adequately, that is enough. Even if they do care, they may not understand what the SWL or DXer wants. Many SWLs have reported reception to some domestic SW stations and have received long, eloquent letters in return—letters which, unfortunately, make no reference whatsoever to verifying reception or even that the listener heard them at all. The language problem hits the SWL/DXer from two angles. The SWL/DXer is faced with the task of understanding enough of the program to convince the station that the reception indeed took place. And then the listener has to compose a reception report in a language that the station personnel can understand. While SWLs can write in English to any major international broadcaster with no worries, such is not the case when a letter is sent to a station in an isolated part of South America or Asia. (Think about it for a second: How quickly, if at all, would you—or could you—answer a letter written in Indonesian?) To help SWLs in this regard, many SWL clubs have put together report "forms" in various foreign languages along with a list of words and phrases which can be inserted as necessary into the form to describe the reception.

Another problem is the postal services of many countries. Theft of mail is not uncommon in some countries, particularly of letters from the United States (which are often thought to contain money). Commemorative stamps on an envelope may cause it to become "lost," and stamps depicting the American flag can produce a hostile reaction in some countries. Addresses can be a problem as well. While simple addresses suffice for most international broadcasters, domestic stations are far less well-known, and an improperly or incompletely addressed letter may be delivered to the wrong station or not delivered at all.

The final problem faced in QSLing domestic stations is their resources and personnel. Many are "shoestring" operations and simply cannot spare the money or people to answer SWL reports. It is standard practice to include some form of return postage, either through mint postage stamps of that country or an International Reply Coupon (IRC), with reports to most domestic SW broadcasters. Often reports will "sit" at a station until a particular person decides to answer them. Names of such "verie signers" are given in the QSL columns of various SWL club bulletins. To catch the attention of station personnel, some SWLs include souvenirs such as postage stamps and picture postcards with their reports.

In some cases, a prepared reply card is sent with another report if previous reports fail to produce a QSL. Such a prepared card contains all confirming data and merely requires station personnel to sign and mail it.

Why do some SWLs and DXers go to such trouble? For one thing, it is often exciting to receive a letter with exotic stamps and an unusual postmark. A letter or QSL card in a strange language has more appeal for them than a bland QSL card sent out by the thousands from an international broadcaster. Perhaps most importantly, such QSLs usually have a far more personal touch. Letters are true personal letters rather than a form, and printed QSL cards often have handwritten notes on them. Sometimes a gift or souvenir may be enclosed. Cancelled postage stamps of the country may be received or a station sticker or decal may be sent. For years, many Latin American stations would send colorful station pennants (similar to Figure 1-3) to SWLs. Unfortunately, economic considerations have forced many stations to curtail this practice. However, the special and genuine interest many domestic SW stations show toward SWL letters and reports still continues. In fact, many international broadcasters would do well to display a similar attitude instead of regarding SWL letters and reports as a bothersome chore.

In the following sections, we'll look at some domestic SW stations around the world. Although the information is as accurate as possible at the time this book is written, it is possible (quite likely, in fact) that some changes will have taken place by the time you read this. For the latest details, the latest copy of *The World Radio TV Handbook* or SWL club bulletins should be consulted.

EUROPE

It might seem surprising that a highly advanced nation such as West Germany, with its myriad AM, FM, and TV outlets blanketing the entire nation, would support domestic SW broadcasting. Yet it does, with four domestic broadcasters relaying their programs on shortwave. Actually, such stations are probably more listened to in East Germany and by Germans residing or traveling in the rest of Europe than in West Germany itself. These stations can be heard in North America, with SWLs in the eastern portions of the continent finding reception much easier.

Perhaps the easiest such station to receive in North America is Sudwestfunk, an AM and FM broadcasting network headquartered in Baden-Baden. They operate a 20-KW transmitter on 7265 kHz relaying their normal programming. The frequency is unfortunately in the North American 40-meter amateur radio allocation, resulting in heavy QRM from hams using SSB. Sign-on is currently scheduled for around 0400 UTC, and continues until sunrise at the transmitter site. This happens to nicely coincide with the time many hams in eastern North America are going to bed, particularly on weeknights, and often Sudwestfunk can break through the QRM with readable signals. In winter, SWLs along the east coast may be able to hear Sudwestfunk prior to its scheduled sign-off at 2300. The programming, all in German, consists of German and American pop tunes, and you can hear "Baden Baden" in most station identifications.

The other domestic SW stations in West Germany pose a challenge for even the most experienced DXer. One is Sender Freies Berlin, on 6190 kHz, located in Bremen. In addition to its own programs, Sender Freies Berlin also relays the programs of Radio Bremen and a special all-night (that is, 2130 to 0500 UTC) program of pop music and news produced by the West German government. It is during the 2130–0500 UTC period that reception in North America is most likely. However, QRM is often heavy and you may find that reception depends on which frequencies are or are not being used by various international broadcasters. A more challenging target is Suddeutscher Rundfunk on 6030 kHz. This station signs on around 0400 UTC and operates until approximately 2300; however, the frequency is often subjected to heavy QRM, and its 20-KW transmitter often cannot break through the din.

An interesting station is RIAS on 6005 kHz. "RIAS" stands for *R*adio *I*n the *A*merican *S*ector, and is operated by the U.S. Information Agency in the American sector of West Berlin. All programs are in German, and include news, weather reports, and music. As you might expect, the East German population is as much, if not more, of a target audience than the residents of West Berlin. The 6005-kHz transmitter operates with 100 KW on a continuous basis, and the frequency is worth checking whenever propagation between West Berlin and your listening location is possible.

East Germany also relays one of its domestic services, Stimme der DDR, on 6115 kHz with 50 KW. It can be heard in North America from sign-on shortly before 0000 UTC until sunrise at the transmitter site near Berlin. (This frequency is also used by its international service, Radio Berlin International.) Stimme der DDR is also relayed on 9730 kHz, again with 50 KW, from approximately 2315 to 0530 UTC.

Albania's major domestic network is relayed on both 5020 and 5057 kHz by Radio Gjirokaster. The only language you'll hear spoken is Albanian, often with alternating male and female announcers, along with some music that often has a distinctly local flavor. Like Radio Tirana's international service, the audio quality of Radio Gjirokaster tends to be poor and some slight frequency variation can be expected. During winter, listeners in eastern North America can sometimes catch its sign-off just prior to 1930 UTC. Everyone has a shot at its 0400 UTC sign-on. Both frequencies use 50 KW but may not operate simultaneously; it's best to check both.

In the early 1980s, Italy's Radiotelevisione Italiana (RAI) lost its monopoly over broadcasting in Italy due to an unexpected court ruling. The result was that numerous private (and at first unlicensed) broadcasting stations were established on AM and FM. A few of these also set up SW broadcasting operations; the best heard of these was Radio Milano International broadcasting from Milan on 9435 kHz. At the time this book was written, Radio Milano International operated on a

24-hour schedule in Italian with most programs involving Italian and English pop tunes. This situation may have changed, since the Italian government was reported to be developing a system of formal licenses and other controls for pirate broadcasters. However, it does seem that private domestic SW broadcasting will be allowed from Italy in the future.

Bulgaria relays its first home service network, known as "Horizont," continuously on 7670 and 11765 kHz. The latter frequency uses a 250-KW transmitter, while the 7670-kHz outlet is only 15 KW. All programs are in Bulgarian and include plenty of music.

USSR DOMESTIC SHORTWAVE

The Soviet Union is a massive country stretching across two continents—Europe and Asia—with segments of its population living and working in extremely isolated areas. This makes shortwave the most practical method of broadcasting to such citizens, resulting in an extensive network of domestic SW services.

In the previous chapter, we examined how some of the various republics making up the USSR have their own international broadcasting services. In the same manner, domestic SW broadcasting has a strong regional slant. There are three national (or "all union," to use the Soviet term) programming networks in the Soviet Union. The first is known as the "all union" program, and is a combination of news, talk, and music. The second network is called "Mayak," which means "lighthouse" in Russian. Mayak is primarily music, with news on the hour and half-hour; live sports events are sometimes broadcast. Unlike the other national networks, Mayak is broadcast continuously. (The other two generally operate from approximately dawn to midnight, local time; the USSR stretches across eleven time zones.) The third national network is mainly classical music and literary, including readings of poetry and dramatic works.

The various republic broadcasting stations generally transmit two programs of their own. The first consists of the "all union" program with news and various features added at the station. The second program originates in the republic itself and is in the main language spoken in the republic. In addition, several regions and areas also have their own programming with more local items such as weather reports and activities; these are in the language mainly used in the particular area. Finally, major cities such as Moscow and Leningrad have their own local programs, although these are currently broadcast only on AM or FM.

Reception of USSR domestic SW stations in North America centers around two periods. The first is from approximately 2000 to 0000 UTC, and the second begins around 1000 UTC and continues until your local sunrise. These reception times are due to the fact that many USSR domestic SW stations are found on frequencies below 6000 kHz. Unfortunately, trying to determine which station you're listening to is usually difficult. All identifications in Russian consist of the word "govorit" followed by the name of the city where the originating *studio* is, not the station itself. Moreover, available transmitter site data for Soviet broadcasting stations is incomplete and out of date, and QSLs from domestic SW stations are unknown. While you'll quickly learn to recognize a Soviet regional station (the interval signals for the "all union" program and Radio Moscow are identical), trying to determine exactly which one you're hearing is a challenge. The latest edition of *The World Radio TV Handbook* carries the latest data on likely Soviet domestic SW transmitter sites.

Nevertheless, you might want to try 4010, 4040, 4050, 4060, 4780, 4785, 4825, 4930, and 5065 kHz around 2000 to 0000 UTC. From 1000 UTC to your local sunrise, try 4040, 4055, 4395, 4085, and 4610 kHz in addition to the first group of frequencies.

The Soviet Union also relays its domestic services to its naval forces and merchant ship fleet traveling the world's oceans. This is known as Radiostantsiya Rodina (radio station motherland), known by SWLs as "Rodina." This service relays segments of the Mayak service along with special features for seamen and Soviet citizens abroad. "Rodina" is relayed over both domestic SW transmitters as well as the facilities of Radio Moscow.

Some DXers have a special interest in Soviet domestic SW stations. They try to identify the transmitter locations that Soviet authorities won't disclose and keep track of frequency and programming changes. The situation for SWLs is a bit different. Naturally, the USSR domestic services won't be of great interest to you unless you happen to speak Russian or one of the other languages of the USSR. However, it is often interesting to listen to the "style" of USSR domestic services, such as the tone and approach used by the announcers, as well as the musical selections played. And it is striking to contemplate during reception of a Soviet domestic station that it is likely that someone in the Siberian frontier or Arctic north is listening to the same station—an example of how shortwave can bridge distances and allow a glimpse directly into a vastly different culture.

AFRICA

Domestic SW broadcasting from Africa is really lively. There's a large number of stations that can be heard, and almost every one is interesting.

Perhaps the easiest place to start in listening to Africa is South Africa. In addition to Radio RSA, SABC operates four domestic shortwave networks which can be heard in North America with good signals. Two of the networks are in English and the South African language Afrikaans; these are both noncommercial. Another is Radio 5, a commercial pop music network. The fourth is Radio Orion, an all-night (in South Africa) commercial network with most music falling into the so-called "easy listening" category. All of these networks are also carried on AM and FM stations in South Africa. There is another English commercial network and networks in various African languages which are carried only over AM and FM stations. All of these networks, like Radio RSA, are operated by the South African government.

The noncommercial English service can be heard from 0600 UTC on 4835 kHz until sunrise in South Africa. The Afrikaans network can be heard beginning at the same time on 4880 kHz. It can also be heard beginning around 0515 UTC on 3320 and 9560 kHz. Radio Five can be found around 0300 to 0500 UTC on 3250 kHz and from approximately 0600 UTC on 7170 kHz. Radio Orion can be found on most of these frequencies at 2200 to 0400 UTC.

South Africa's controversial "homelands" policy calls for large segments of that nation's black population to be relocated to designated areas, which supposedly will be four independent, self-governing areas. At the present time, no nation but South Africa recognizes these "homelands" as sovereign nations. This has not prevented the establishment of broadcasting facilities in these new "nations." Only one operates on shortwave—Capital Radio, located in the homeland of Transkei. The format is remarkably similar to an American "Top 40" AM station and can be heard beginning at 0300 UTC on 7150 kHz until sunrise in South Africa.

South Africa controls the territory of South-West Africa, which is also known as Namibia. Its domestic SW broadcasting service is known as Radio South-West Africa, and is yet another service of SABC (even their QSL cards are remarkably similar). The best chance to hear this station in North America is from 2200 to 0400 UTC with the all-night service. The format is easy listening and pop music with announcements in English, Afrikaans, and German. Frequencies used during this period include 3270, 3295, 6185, and 7190 kHz, although not all of these may be in use at once. In addition to this service, there are four networks, known as Radio Owambo, Radio Herero, Radio Damara-Nama, and Radio Kavango, which program in various African dialects. The easiest to hear in North America is Radio Herero, which can be heard on 3270 and 7190 kHz beginning at 0400 UTC after the end of the all-night service.

Nigeria also operates several domestic SW stations which broadcast many programs in English; most of these can be heard with much difficulty in North America. The main domestic network is the Federal Radio Corporation of Nigeria (FRCN), which identifies itself on the air as Radio Nigeria. Its stations at the capital of Lagos can be heard on both 3326 and 4990 kHz at 0430 UTC and continuing until sunrise in Nigeria. You can also catch Lagos on 7285 kHz beginning at their 1000 UTC sign-on. (It's best to try on a weeknight when QRM from ham radio operators is less.) Each frequency uses 50-KW transmitters, and most of the programming you'll hear will be in English. In winter, you might want to try for the 3326- and 4990-kHz stations prior to the scheduled sign-off shortly after 2300 UTC.

FRCN also operates regional SW transmitters in other parts of Nigeria. While English is often used, you'll also hear the local languages of the region and some of the distinctive African pop music known as "high life." Perhaps the easiest regional outlet to hear in North America is the one at Kaduna on 4770 kHz. This can be heard at their sign-on at 0400 UTC or before sign-off at 2300 UTC. You'll hear "Kaduna" used in station identifications. A somewhat more

Figure 7.1 The desolate landscape of Nambia is shown on this QSL from Radio SWA.

challenging target is the station at Ilorin, the capital of the Kwara state. It can be heard on 7145 kHz from sign-on around 0345 UTC; you'll face heavy QRM from ham radio operators in this range, and the transmitter is rated at only 10 KW. While government operated, this station is administratively separate from the FRCN and identifies itself as "Radio Kwara" or the "Kwara State Broadcasting Service." You may also be able to hear before its scheduled sign-off at 2300 UTC.

The remaining Nigerian domestic SW stations present a greater challenge, mainly because they operate in the crowded 49-meter international broadcasting band. All can produce readable signals in North America *if* the QRM doesn't cover their signals. Most sign on at 0400 or 0430 UTC, and reception is possible until sunrise in Nigeria; a second time to try for them, particularly in winter, is from 2200 until their scheduled sign-off around 2300 UTC. Of particular interest is the station at Enugu on 6025 kHz; this was the capital of the breakaway nation of Biafra during Nigeria's bloody civil war and for a brief time operated the Voice of Biafra. The Voice of Biafra vanished with the defeat of the Biafran rebels (primarily members of the Ibo tribe), but while it lasted it allowed SWLs to literally hear a revolution in progress and to get a brief glimpse into the problem of tribalism in Africa, a concept which is seldom mentioned on African broadcasters and is poorly understood in the western world. Other regional outlets you can aim for include Jos on 5965 kHz, Ibadan on 6050 kHz, Kaduna on 6090 and

6175 kHz, Maiduguri on 6140 kHz, Calabar on 6145 kHz, and Sokoto on 6195 kHz.

Another African nation which uses plenty of English in its domestic SW broadcasting is Botswana. They also have one of the most unusual interval signals you'll ever hear—cow bells and the sound of farm animals. You can hear Radio Botswana at its 0400 UTC sign on on 3356, 4848, and 5965 kHz; it can also be heard on 7255 kHz from 0500 to 0630 UTC if QRM permits. In addition to English, you'll hear various African dialects used. If you're a QSL collector, you're out of luck with Radio Botswana—they normally refuse to verify listener reports. The only exception appears for reports about truly exceptional circumstances; one DXer did receive a QSL letter in response to a report about off-frequency operation by one of Radio Botswana's transmitters.

Swaziland is another African nation where a large amount of English is spoken. Most North American SWLs hear Swaziland via the Trans World Radio station there. The government-operated Swaziland Broadcasting Service currently operates only on AM and FM. However, there is a commercial SW broadcasting station known as Swaziland Commercial Radio. This station is backed by South African money and broadcasts in a variety of languages, including some used in India. You can try for it on 6155 and 9704 kHz from its scheduled 0500 UTC sign-on. The station identification is "Radio SR."

Lesotho is a tiny nation completely surrounded by South Africa, and much English is also spoken there. Presently the government-operated Radio Lesotho can be heard on 4800 kHz from its 0300 UTC sign-on until sunrise in Lesotho; you'll probably find best reception after 0500 UTC when QRM from Latin American stations also on 4800 kHz is less. Programming involves both local and American music as well as newscasts and frequent time checks.

Since France was a major colonial power in Africa, it's not surprising that you'll hear plenty of French on domestic SW stations in Africa. In countries where numerous local languages are spoken, French may often be the common language for government and business. One such country is Togo, whose national broadcasting service, Radiodiffusion Television Togolaise, has a French name. It can be heard on 5047 kHz, where it operates a 100-KW transmitter, at its sign-on around 0530 UTC or before its sign-off at 0000 UTC. (The frequency may vary slightly.) Listen for its French identification as "Ici Lome," which is the nation's capital and where the studios are located. Togo also operates a separate station at Kara, which can be heard on 3222 kHz before its 2300 UTC sign-off or at its 0530 UTC sign-on. Kara carries separate programming from the 5047-kHz station and identifies as "Ici radiodiffusion de Kara." In addition to French, you'll also hear some exotic local languages used by both stations.

A station new SWLs sometimes confuse with Togo is Radio Centrafrique, from the Central African Republic, on 5035 kHz. Like Togo, this frequency uses a powerful 100-KW transmitter but sometimes drifts off its assigned frequency. Listen carefully for its station identification of "Ici Radio Centrafrique" and its interval signal of a piano. It can be heard at its sign-on at 0430 UTC and is often heard very well during winter in eastern North America before its 2300 UTC sign-off.

"Burkina Faso" may sound like the name of a character from a James Bond novel, but it's actually the new name of the nation formerly known as Upper Volta. Its broadcasting service, Radiodiffusion Television Burkina, can be heard on 4815 kHz at its 0530 UTC sign-on and before its 0000 UTC sign-off. You'll hear both French and "high life" music along with some lengthy talks.

An interesting broadcasting situation exists in Gabon. It is home to a commercial broadcaster known as Africa Number 1, which operates on numerous frequencies with 250- and 500-KW transmitters. Its income is derived primarily from serving as a relay base for Radio Japan and Radio France International as well as selling air time to various religious groups. In many respects, Africa Number 1 is similar to Radio Trans Europe in Portugal. If you spend much time tuning the international SW broadcasting bands, you'll eventually come across this station. Why not try for Gabon's government-operated domestic SW service, Radiodiffusion Television Gabonaise? It can be heard in French signing on at 0630 UTC on 4770 and 7270 kHz; it can also be heard until 0000 UTC on the former frequency. You're likely to hear "high life" and other pop music. Listen for station identifications beginning with "Ici Libreville," which is the nation's capital.

Another African station you'll hear plenty of French from is Radiodiffusion Television Ivoirienne, in the nation of Ivory Coast. Check 4940 and 7215 kHz at their scheduled 0600 UTC sign-on. French is also used along with local languages by Guinea's Radiodiffusion Nationale, which can be heard on 4910, 6155, and 7125 kHz at 0600 UTC sign-on. Listen for the slogan "la voix de la revolution." In the past, this station often ran lengthy political harangues. Now you're more likely to hear "high life" music and tribal rhythmns.

Like Botswana, some African domestic stations have unusual interval signals. Ever heard a "cora?" That's an instrument native to the nation of Senegal, and is used at sign-on of Radiodiffusion Television Senegal. Tune for it at 0600 UTC; you'll hear French, Arabic, and local languages used. Or how about a "tam-tam?" That's the interval signal for the voice of the nation of Benin, Office de Radiodiffusion et Television du Benin. Try for them on 4870 and 5025 kHz at 0400 UTC sign-on; the latter is a regional program for the Parakou area of Benin. In addition to French, you'll hear some English used as well; African and Western pop music is featured. Listen for the station identification of "Ici la voix de la revolution Beninoise."

In the northern part of Africa, Arabic replaces French as the language you're most likely to hear. However, some French influence remains present among stations located in former French colonies and territory. An example is Tunisia, whose national broadcasting service is known as Radio Television Tunisienne. However, their French service is carried only on AM and FM stations. You can hear their Arabic home service on SW easily, though. It is currently scheduled from 0430 to 0600 UTC on 7225 and 9675 kHz and afternoons until their scheduled sign-off at 2330 UTC on 11730 and 15225 kHz. The programming is very different from what you hear from stations in southern Africa; traditional Arabic music and recitations from the Holy Quran are more likely programming fare.

A nation whose Arabic and French networks are relayed on shortwave is Algeria. Radiodiffusion Television Algerienne can be heard in Arabic in afternoons until 2100 UTC on 11715 kHz and until 2200 UTC on 15370 kHz. You can also hear their sign-on at 0600 UTC on both 6145 and 6160 kHz. Their French network can be heard until 2100 UTC on 9685 kHz or beginning at sign-on at 0500 UTC until sunrise in Algeria on 9685 kHz. At one time Algeria requested that all reception reports, even for its Arabic programs, be written in French.

Portugal also had colonial interests in Africa, and thus Portuguese is often heard on domestic stations located in their former colonies. One example is Angola. Radio Nacional de Angola, where Portuguese is the language of government and commerce and over ten different local vernaculars are spoken in everyday use. One network is devoted exclusively to Portuguese programming and can be heard on 7215 and 9660 kHz from 0530 UTC sign-on. It may also be heard on 3355 and 4820 kHz until sign-off at 0530 UTC. During the first few months of Angola's independence, this network ran what seemed to be live political speeches, complete with excited cheering and crowd shouts. The programming is considerably quieter now, with much Western music and sedate announcers. There is another Radio Nacional de Angola network which broadcasts in French, English, Spanish, and the various local languages. It can be heard on 7245, 9535, and 11955 kHz at its 0530 UTC sign-on; the first two frequencies can also be heard before sign-off at 2300 UTC.

Angola also has several regional SW stations whose programs are independent of the various national networks. Each station's name begins with "emissor regional do" followed by the station's location. All of these stations are low-powered, with one station using 25 KW, three using 10 KW, and the rest using only 5 or 1 KW of transmitter power. Needless to say, these stations are favorites with DXers. The easiest to hear is the station at Huila, with 25 KW, on 4B20 kHz; it can be heard at its sign-on at 0400 UTC. (Be careful not to confuse this with the Portuguese outlet of Radio Nacional de Angola on this frequency as well.) Another likely target is Huambo on 5060 and 7160 kHz at its 0400 UTC sign-on. A station that stands out due to its oddball frequency is Moxico on 5192 kHz; it is scheduled to sign on at 0500 UTC and to sign off at 2200 UTC. The other Angolan regionals are more difficult targets, although all are possible to receive in some part of North America depending upon QRM.

Another former Portuguese colony is Mozambique. Years ago, broadcasting in Mozambique was the responsibility of Radio Clube do Mozambique, and it was largely controlled by South African interests. One service, known as Lourenco Marques Radio, was an English commercial service intended for South Africa. With independence, all broadcasting in Mozambique became a government operation. Radio Mozambique is the national broadcasting organization of that country, and its main Portuguese network is known as Emissao Nacional. It can be heard on 3210, 4865, and 6115 kHz from its 0300 UTC sign-on until sunrise in Mozambique. Like Angola, Mozambique has several regional SW stations. The easiest to hear is Emissao Provincial de Sofala, which can be heard on 6090 and 9667 kHz in local languages beginning at 0300 UTC sign-on until sunrise in Mozambique. As is the case with Angola, the remaining regional stations in Mozambique are challenging DX targets.

THE NEAR AND MIDDLE EAST

Arab language and music, along with recitations from the Quran, make up the bulk of the domestic programming from this area. A good example is the domestic programming of the Broadcasting Service of the Kingdom of Saudi Arabia. You can hear them on 5875, 7220, 7290, 9720, and 11950 kHz at their 0300 sign-on; transmitter powers of up to 350 KW are used and signals are usually good. Saudi Arabia also broadcasts programs consisting exclusively of recitations from the Quran beginning from 0500 to 0700 UTC on 11730 kHz and at 1900 to 2100 UTC on 7270 and 9655 kHz.

The programming from Radio Kuwait is a bit more Western. You can even hear Western pop music on their Arabic programs. These can be heard afternoons on 9840, 9880, and 11990 kHz until sign-off at 2215 UTC. Their domestic English service is also relayed on 11675 kHz at 1800 to 2100 UTC. This latter service includes both Arabic and Western music along with newscasts concentrating on Kuwait and Middle East events.

Arabic is not the major language of every Islamic nation. The language of Turkey is, appropriately enough, Turkish. Turkey is home to a major international broadcaster, the Voice of Turkey, which broadcasts in English to North America and is quite easy to hear. It is also home to two domestic SW stations which are superb DX challenges in North America. One is Turkish Police Radio on 6340 kHz with only 1 KW of power. It is operated by the Turkish police to improve relations between the police and citizens, and programs consist of Turkish music and announcements. The best time to try for it is at 0600 UTC sign-on. The second station is operated by the Turkish State Meteorological Service on 6900 kHz with 2.5 KW of power. Programming is music and, naturally enough, weather reports. It signs on at 0400 UTC, but QRM on this frequency from utility stations is often severe.

THE INDIAN SUBCONTINENT

Indian domestic SW stations are invariably DX for North American listeners. The propagational path at any time or on any frequency is tough, and it takes patience and a little help from the ionosphere to hear them. However, the exotic music and languages you can hear make the effort required worthwhile.

Broadcasting in India is the responsibility of the government's All India

Radio. The domestic SW service is just as varied and complex as the nation itself. Over twenty different domestic SW stations use dozens of different frequencies. QSL collectors are pleased by the fact that most of these stations will verify reception reports. The programming could be in any of the languages spoken in India, such as Bengali, Hindi, Punjabi, Sanskrit, Tamil, Urdu, and English. The music will usually be of the type known among SWLs as "subcontinent music," which consists mainly of stringed instruments and flutes.

Most Indian reception in North America is the result of gray-line propagation, meaning that you should try for such stations a half-hour before and after your local sunset and sunrise. Among the stations you might try for are Gauhati on 4775 kHz, Calcutta on 4820 kHz, Bombay on 4840 kHz, Delhi on 4860 kHz, Madras on 4920 kHz, and Aizawal on 5050 kHz.

Pakistan has a more modest domestic SW network, with only five stations. However, they may be even more challenging targets than the Indian stations. Outlets you might want to try your luck on include Karachi on 4815 kHz, Ouetta on 4879 kHz, and Peshawar on 4950 kHz. The main language of Pakistan is Urdu, but you likely won't be able to tell much difference between its programming and that of an All India Radio outlet.

As noted in the previous chapter, Sri Lanka is home to relay stations operated by the Voice of America and Trans World Radio. The government broadcasting service is the Sri Lanka Broadcasting Corporation, and it runs a modest international broadcasting operation. Much more interesting, and difficult to hear, are the various domestic SW services. 4902 kHz is used exclusively for programs in the Sinhala language, while 5020 kHz is used for Tamil programs. An English service is aired on 6130 kHz, although reception of it in North America is quite difficult.

THE FAR EAST AND ASIA

The nations of Asia are diverse. They range from such highly advanced, technological countries as Japan to underdeveloped states still locked in the past. Domestic SW broadcasters in Asia reflect this diversity.

Japan would seem at first to be an unlikely spot for domestic SW broadcasting, since it is every bit as heavily saturated with AM, FM, and TV stations as the United States and Canada. But it is home to the Nihon Shortwave Broadcasting Company, which operates Radio Tanpa. This is a commercial broadcaster with all programs in Japanese, although some of the commercials and music are in English. Reception in North America is more challenging than that of Radio Japan, although it is not very difficult. Two separate programs are transmitted by Radio Tanpa. The first operates on an almost 24-hour schedule on 3925, 6055, and 9595 kHz. The second operates from 2300 to 0900 UTC on 3945, 6115, and 9760 kHz. You'll probably find reception best beginning at a couple of hours before your local sunrise.

The Far East and Asia   141

Unlike Japan, Burma is a closed, inward-looking state that has little in the way of Western technology. Moreover, its Burma Broadcasting Service is more difficult to hear than Radio Tanpa. Best reception is usually found on 5985 kHz at 0930 UTC sign-on; Burmese is the language used. Listeners in western North America may be able to hear the start of their English programs after 1430 UTC. Another possible frequency is 4725 kHz, where programs in such regional languages as Chin, Kayah, Moh, and Shan are transmitted, beginning with 1030 UTC sign-on. Burma can be a challenge for the QSL collector; some SWLs have even resorted to sending their reports to the Burmese embassy in Washington, DC.

Like the Soviet Union, China is a vast nation with some citizens located in isolated areas and it relies heavily on shortwave to reach them. Unlike the USSR, however, China makes available up-to-date information about its domestic SW stations, and most of them issue QSLs in response to reception reports. As with Radio Beijing, these domestic stations have mirrored the changes within China itself. The shrill, strident revolutionary operas and political monologues have been replaced with traditional Chinese music and opera along with some surprising injections of Western culture, such as Linda Ronstadt songs.

China's domestic SW stations are numerous and range from relatively easy SWL targets to prime DX. For example, the Fujian province stations on 5770 and 7025 kHz are often heard well in North America from 1000 UTC until local sunrise. But try hearing the station at Kunming, Yunnan on 2310 kHz around 1230 UTC.

SWLs and DXers are aided by the oddball frequencies used by many Chinese domestic stations. For example, the station at Urumqui, Xinjiang stands out among the utilities on 4220 kHz; another station at the same location operates on 4330 kHz. Another station on 4330 kHz is at Haixia, Fujian; try for this one beginning at its sign-on at 1200 UTC when the Urumqui station is off the air. Other unusual frequencies include 3400, 3535, 3640, 4035, 4045, 4250, 4380, 4460, 5125, 5163, 5170, 5240, 5265, 5320, 5440, 5770, 6400, 6493, 6500, 6665, 6750, 6765, 6790, 6937, 6974, 7504, 7516, 7525, 7770, 7850, 7935, 8007, 8667, 10245, and 11040 kHz. China also uses more conventional frequencies; in fact, perhaps the best signal in eastern North America during winter is found around 1100 UTC on 4840 kHz from the station located at Haixia, Fujian.

China also has a "domestic" network intended for reception in Taiwan instead of an "international" service. This reflects the Chinese position that Taiwan is not a separate nation but rather just another province of China. Broadcasting in Taiwan itself is under the auspices of the Broadcasting Corporation of China (BCC), which describes itself as a "private enterprise under government contract" (or, more precisely, "control"). Of the domestic stations intended for reception on Taiwan itself, only one operates on SW. This is the station at Hsinchu on 3215 kHz, which might be audible shortly before your local sunrise. Far easier to hear are the stations operated by the Central Broadcasting System, a BCC subsidiary which provides a "domestic" service for the Chinese mainland.

It is scheduled from 2155 to 0955 UTC on 7105 kHz; at 0950 UTC it begins operations on 7250 kHz. You may also be able to hear it at this latter time on 11775 and 11905 kHz. These are actually two separate networks. Both program entirely in Chinese, with plenty of talk and music.

Vietnam is a nation that still looms large in the American consciousness. Its external service, the Voice of Vietnam, was widely quoted in American news media and was often the first source of information about positions taken by the Hanoi government. (The American press also managed to continuously misidentify this station as "Radio Hanoi"—a name the station never used, leaving alert SWLs wondering which station, if any, various news reporters were actually listening to.) The Voice of Vietnam is still in operation today, and there is an extensive domestic SW broadcasting network as well. The main station at Hanoi operates on 6450, 7420 and 10060 kHz until approximately 1600 UTC; these may be possible at your location in the dawn hours. Several other domestic SW stations are found in Vietnam, most being prime DX targets. Fortunately, many operate on unusual frequencies (5600 and 6822 kHz, for example) and stand out in the band when reception is possible. All programming is in Vietnamese or Montagnard (a language spoken by a mountain people within the borders of Vietnam who have never accepted Vietnamese rule). All frequencies used by Vietnamese stations are highly variable.

Malaysia is a federation made up of a peninsula and part of an island. It's not surprising that domestic SW is used as part of the government's Radio Television Malaysia broadcasting system. The major languages of this nation are Malaysian, Chinese, and English, and you'll hear all three used by Radio Malaysia. The basic Malaysian home service is relayed continuously on 5965 kHz and until 0900 UTC on 7200 kHz. It may also be audible around your local sunrise on both 9515 and 9710 kHz. The English home service was scheduled, at the time this book was written, at 0900 to 1600 UTC on 7295 kHz.

Malaysia is divided into districts, and two have their own separate government-operated broadcasting systems. The Sabah district operates Radio Malaysia Kota Kinabalu. It has one network devoted entirely to English and Malay programs 4970 and 5980 kHz; it signs on at 0900 UTC, and English is scheduled beginning at 1130 UTC. Your local sunrise is the best time to try for these two frequencies. The Sarawak district operates Radio Malaysia Sarawak. Their Chinese and English network on 4950 and 7160 kHz has a scheduled 0030 UTC sign-on; again, your local sunrise is a good time to try for these.

Singapore was once a part of Malaysia. It's now an independent state and a rising economic power. The BBC maintains a relay station there, but it's not much more difficult to hear the Singapore Broadcasting Corporation's domestic SW service. Its English network, known as Radio One, can be heard on 5052 and 11940 kHz around your local dawn. Don't expect much in the way of exotic Oriental music on these frequencies; the programming tends to be Western music, including rock.

Indonesia is composed of the numerous islands making up the Indonesian archipelago. It may very well be the favorite target of DXers worldwide. Indonesia has numerous private, commercial stations on the AM band, but domestic SW broadcasting is restricted to the government's Radio Republik Indonesia (RRI). RRI's domestic SW network is massive; over 100 different frequencies are used by more than 70 different stations. Most frequencies are found in the tropical broadcasting bands, and transmitter powers of 10 KW or less are generally used. The various RRI stations also tend to be excellent verifiers of reception reports, often with long personal letters in Indonesian. (To facilitate writing reception reports, form letters in Indonesian have been developed and are available from some SWL clubs.) But the DX challenge and readiness to issue QSLs does not fully explain the popularity of Indonesian stations among DXers. Part of the reason may lie in the exotic nature of the programming itself. Indonesian is a truly unusual language to Western ears, and many SWLs find a lilting, almost musical quality to it. RRI stations also feature plenty of authentic "South Seas"- type music. Finally, there is the distinctive interval signal used by all RRI stations called "Love Ambon," also known as the "Song of the Coconut Islands." This is played just before the hour (using organ, piano, flute, and vibes) and is followed by three time "pips" and a newscast in Indonesian.

Indonesian DXing, including positively identifying the stations you do hear, is an advanced listening activity that requires time and patience to master. However, there are several RRI domestic stations which can be heard throughout North America with a little effort. Among these are Dili on 3120 and 3307 kHz, Nanjarmasin on 3250 kHz, Ternate on 3345 kHz, Tanjungkarang on 3395 kHz, Celebes on 4720 kHz, Sumatra on 4764 kHz, Jakarta on 4774 and 7270 kHz, Ambon on 4845 kHz, and Jayapura on 9615 kHz. Your best chances for hearing these stations is around your local dawn. Don't be too surprised if you find yourself becoming one of those DXers "hooked" on Indonesia.

AUSTRALIA AND THE PACIFIC

Broadcasting in Australia is conducted by private commercial interests as well as the Australian Broadcasting Corporation (ABC). The ABC is responsible for three SW stations intended for reception in the nation's vast and sparsely inhabited interior. All programs are relays of the various ABC regional networks, which are broadcast on AM and FM as well as SW. Thus, all programs you hear will be in English (albeit the Australian version of it). The identification announcement is usually a simple "This is the ABC." And all three stations will verify reception reports as well.

If you're a SWL in eastern North America, the ABC station at Perth may well be the most physically distant station you can hear on shortwave. The transmitter is located at Wanneroo, near Perth, on Australia's west coast and can often be heard with good signals throughout North America on 6140 and 9610 kHz from shortly before local sunrise until scheduled sign-off at 1600 UTC. In late evenings and at night during periods of high MUFs, this station can also be heard on 15425 kHz.

Another ABC domestic SW station is located at Lyndhurst, near Melbourne. This station can be heard on 6150 and 9680 kHz during the same time Perth is audible. The third ABC SW outlet is at Brisbane, on the eastern coast, and operates on 4920 and 9660 kHz during the same time period.

The nation of Papua New Guinea is a former Australian territory. While English is the common language of government and business, other languages, such as Melanesian, Pidgin, and Hiri Motu, are spoken. Broadcasting is the responsibility of the National Broadcasting Commission of Papua New Guinea. Since the nation is large and relatively undeveloped, shortwave plays a major role in its broadcasting efforts. Currently, there are over 20 domestic SW stations in operation. A few are relatively easy to hear in North America, while most present challenging DX targets. The easiest one to try for is the National Service outlet on 4890 kHz from your local dawn until scheduled sign-off at 1400 UTC. Usually this frequency puts a good signal into all parts of North America. Programs here will be in English, Pidgin, and Hiri Motu, and the station will verify correct reports. While this is relatively easy DX, the same cannot be said for other Papua New Guinea stations such as Radio Simbu, which operates on 2376 kHz with only 2 KW of power. There are also several stations operating in the 90-meter tropical broadcasting band which are eagerly sought by DXers throughout North America.

Ask people to name an island paradise in the Pacific, and they're likely to mention Tahiti. Tahiti is a part of French Polynesia, and the principal languages spoken are French and Tahitian. Broadcasting there is conducted by the Societe Nationale de Radio Television Francaise d'Outre Mer (usually simply abbreviated RFO), a French governmental body. Fortunately for SWLs, they operate on four different SW frequencies which can be heard well throughout North America. The most reliable are 11825 and 15170 kHz, with best reception usually found after 0100 UTC. French is scheduled until 0300 UTC, followed by Tahitian until 0600 UTC, and a mixture of the two until sign-off at 0730 UTC. During the Tahitian segments, you can hear some of the traditional music associated with the image of a tropical paradise, such as flutes and guitars. During the French programs, however, you will hear French pop tunes and even disco music (proving, perhaps, that paradise isn't what it used to be). This station has traditionally verified reports with colorful and interesting QSL cards; one used a few years ago featured a topless mermaid.

Another French possession in the Pacific is New Caledonia. This has been the scene of recent conflicts between those residents who wish to remain affiliated with France and those who want the island to become an independent state. RFO is likewise responsible for broadcasting in New Caledonia, and uses three SW frequencies. The most widely heard is 7170 kHz, which is often heard beginning around 0700 UTC until sign-off shortly after 1100 UTC. All programs are in

French, and the music is mainly French pops with some English tunes as well.

The Solomon Islands are located just east of Papua New Guinea, and the principal languages are Pidgin and English. The government operates the Solomon Islands Broadcasting Corporation, whose station identification in English is "Radio Happy Isles." Programs alternate between English and Pidgin segments, and can often be heard with good signals until 0730 UTC on 9545 kHz or until 1300 UTC on 5020 kHz.

CANADA

Canada, like Japan, would seem to be an unlikely spot to find domestic SW broadcasting. However, the northern regions of Canada include some of the most rugged and isolated areas of North America, and many persons living there often depend upon SW for news and entertainment from the outside world.

The Canadian Broadcasting Corporation operates two domestic SW outlets that are superb challenges for DXers. Both operate on 6160 kHz, and they are "flea power" outlets—one uses 500 W of power, while the other puts out only 300 W. One is CKZN, in St. John's, Newfoundland, the 300-W station, which relays AM station CBN located in the same city. The other station is CKZU, in Vancouver, British Columbia, which relays Vancouver AM station CBU. Both AM stations operate exclusively in English. CKZN is scheduled at 0930 to 0500 UTC, while CKZU currently operates from 1400 to 0905 UTC. Since both stations operate on the same frequency, it is easy to confuse the two—provided that you can hear them in the first place. As you might expect, your biggest challenge here will be QRM from international SW broadcasters.

There are also five private, commercial AM stations which have SW relays, all operating in English and on a 24-hour schedule. The easiest of the group to hear is CFRX, 6070 kHz, in Toronto, which relays AM station CFRB. Try for this one whenever there is possibly a propagation path between you and Toronto (including midday if you're within a few hundred miles of Toronto). The transmitter power is 1 KW, but it is sufficient to deliver a good signal to most parts of North America. The second easiest target is probably CHNX in Halifax, Nova Scotia, on 6130 kHz. This station relays station CHNS and is often heard in eastern North America in early morning or at sunrise when QRM on the frequency is reduced. It uses 500 W of power. Another station often heard in eastern North America around the same time is CFCX, in Montreal, on 6005 kHz relaying CFCF. This station is another 500-W outlet.

The other private Canadian broadcasters are much more formidable DX challenges. They are CFVP, Calgary, Alberta, on 6030 kHz (using 100 W) and CKFX, Vancouver, on 6080 kHz. The transmitter power of this latter station is only 10 W! (Don't give up hope—it has been heard in eastern North America when the frequency is clear.)

LATIN AMERICA

For the purposes of this section, "Latin America" is defined as all of Central and South America plus the Caribbean region. This definition isn't precisely accurate from a geographical or cultural perspective, but it does make sense from the standpoint of radio propagation and reception.

Not everything you hear from "south of the border" will be in Spanish. For example, the major language of the Central American nation of Belize is English. This nation was formerly known as British Honduras, and its capital is Belize City. You can hear Radio Belize on 3285 kHz until scheduled sign-off at 0510 UTC; the 2200 to 0300 UTC period is usually devoted to Spanish programs. Radio Belize can often be heard at their 1100 UTC sign-on in English. An English voice from South America is the Guyana Broadcasting Corporation on 5950 kHz. It has a scheduled sign-on at approximately 0730 UTC and is often heard in North America in the 0830 to 1000 UTC period with an eclectic format, including a variety of music (including religious), local announcements of funerals and birthdays, and frequent time checks.

Most people in North America had never heard of the Falkland Islands until the 1982 war between Great Britain and Argentina over them. But DXers throughout the world had long been familiar with them, since the Falkland Islands Broadcasting Station had ranked as one of the top DX targets since it first started operations. In years past, the station used only a few hundred watts of power and operated on frequencies in the middle of the 75-meter amateur radio band or in the 120-meter band. The combination of low power, a difficult propagation path, and heavy QRM made the station an extreme challenge. In fact, during the mid-1970s only about a dozen DXers in North America had managed to hear and verify the station. However, facilities have been sharply upgraded in recent years, including new 3.5-KW transmitters, and the station is much easier to hear. Not that it is easy by any means—now it is simply difficult rather than almost impossible. Frequencies now used include 2380 and 3958 kHz, with sign-on times scheduled for around 0900 UTC. However, both frequencies may not sign on simultaneously, and one frequency may not be used for several weeks. The latest schedules can be found in SWL club bulletins.

English may sometimes be heard over a station that normally broadcasts in Spanish. One example is HRVC, La Voz Evangelica, in Tegucigalpa, Honduras on 4820 kHz. As you might surmise from its name, the station is an evangelical Christian station owned by the Conservative Baptist Home Mission Society in Kansas. In the 0300 to 0600 UTC period, this station broadcasts English religious programs, many of which are aired on AM and FM religious broadcasters in the United States. English may also be heard at times, usually the evening hours, over two stations in Costa Rica: Radio Casino in Limon on 5954 kHz and Faro del Caribe in San Jose on 5055, 6175, and 9645 kHz.

A station that also broadcasts in English, along with French and Creole, is 4VEH, located in Cap Haitien, Haiti. This is another evangelical station operated by American religious organizations. At the time this book was written, it was operating on 4930 kHz, although its assigned frequency is 4950 kHz. At 1100 UTC, they are scheduled to sign on in English and can often be heard well in eastern North America. They can also be heard until 0300 UTC sign-off with religious programs in Creole and French.

French is the major language of French Guiana, a French territory on the northern coast of South America. French Guiana was the site of the notorious Devil's Island prison camp, and today is the launching site used by the French Arienne satellite-carrying rocket. Radio France Internationale also maintains a major relay station in French Guiana, but a more challenging target is the RFO station at Cayenne. The best reception of this station seems to be on both 3385 and 5055 kHz from 0900 UTC sign-on; all programming is in French, and French pop music is played.

But such languages as French or English are seldom heard from Latin America. The vast majority of stations will be in Spanish, with Portuguese (from Brazil) a distant second. You can also find a few stations with programs in local Indian languages such as Quechua, but Spanish will easily dominate over all others.

Fortunately for English-speaking SWLs, Spanish is an easy language to deal with. Many words and phrases bear close similarity to their English equivalents. With a little practice, you'll be able to recognize station names, identification announcements, and even some commercials. As mentioned earlier, several SWL clubs also make Spanish form letters available for reception reports, and Latin American stations tend to be among the friendliest of verifiers.

One thing you'll quickly notice is that there is no such thing as "Latin American music." The music is just as varied as the nations making up the continent. Marimba music will usually indicate that you're listening to a station in Guatemala or a neighboring country. You'll also eventually run across some music best described as "sad flutes"; this indicates you're tuned to a station in Peru, Ecuador, or other nation through which the Andes cross. And the influence of American and British rock is also apparent; you're just as likely to hear Michael Jackson over a Latin American station as you are over a station in Los Angeles.

You might be lucky enough to catch a baseball ("beisbol") game in Spanish via shortwave. More likely, however, is reception of a soccer match. Soccer has a fanatical following in Latin America, and the announcers reflect the intensity in their presentation. The commentary moves along at a pace that can elude even the most determined nonnative speaker; the letter "r" is rolled with great gusto. When a team scores, the word "gol" (goal) is stretched out for what seems to be an eternity.

Although almost all Latin American stations have been assigned call letters, don't expect to hear them used on the air. Invariably, you'll hear a slogan such as "Radio Altiplano," "Ecos del Torbes," and "La Voz de Tolima" used instead to identify the station. This actually works in favor of the SWL who doesn't speak Spanish, since a slogan or name is much easier to catch than individual letters of the alphabet.

The obvious place to start in any examination of Latin America is with

Mexico. One surprising aspect of Mexican domestic SW broadcasting is how few stations such a relatively large nation has; many smaller Latin American countries have far more. Part of this is due to the large number of AM stations found there; several hundred AM stations operate in Mexico, and there are very few areas where several AM stations cannot be heard easily. Another factor may be the influence of American broadcasting and receiver manufacturing patterns; since so much broadcasting and receiving equipment in the formative stages of Mexican broadcasting was imported from the United States, it's not too surprising that broadcasting and listening habits followed the American model.

The easiest private SW outlet to hear is XEWW, La Voz de America Latina, in Mexico City. This is a relay of AM station XEW on 900 kHz (where it can often be heard at night in the southern half of the United States). Its current schedule is 1200 to 0600 UTC on 6165, 9515, and 15160 kHz.

Mexico is also the home of the first 120-meter band station most North American SWLs hear. This is XEJN, Radio Huayacocotla, in the Mexican state of Veracruz, on 2390 kHz. Try for this one shortly after your local sunset. Other domestic SW targets include XEQK, La Hora Exacta, Mexico City, on 9555 kHz with frequent time pips, XEUJ, Radio XEUJ, in Linares, on 5982 kHz, and XEQM, Su Pantera, in Merida, on 6105 KHz. The best times to try for these are often early mornings after 1100 UTC when propagation is possible from Mexico and QRM from international broadcasters is reduced.

Guatemala is relatively easy to hear, thanks to TGNA, Radio Cultural, in Guatemala City, on 3300 and 5955 kHz. This is another evangelical Christian station, and English is currently scheduled at 0245 to 0430 UTC on both frequencies. Other Guatemalan stations that can be heard in North America with some effort include TGBA, Radio Maya de Barillas, in Huehuetenango, on 3325 kHz and TGVN, La Voz de Nahuala, in Nahuala, on 3360 kHz. The best time to try for these latter two stations is around 1100 UTC.

Honduras is easy to hear via the previously mentioned HRVC, La Voz Evangelica, on 4820 kHz. A more challenging target you might want to try for is HRCP, Radio Luz y Vida, in Santa Luis, on 3250 kHz. This is another religious broadcaster, but the programming is entirely in Spanish. They can be heard most evenings until their scheduled 0400 UTC sign-off; you may also be able to hear them on winter mornings at their scheduled 1230 UTC sign-on.

Costa Rica has more domestic SW broadcasters than those mentioned earlier in this chapter. One that is usually heard well throughout North America is TIHB, Radio Reloj, in San Jose, on 4832 kHz. This station operates continuously and is often clear of QRM after 0600 UTC. Other Costa Rican stations which are strong throughout the evening hours include TILX, Radio Columbia, in San Jose on 4825 kHz and TIRS, Radio Impacto, in San Pedro Montes de Oca, on 6150 kHz.

Since the assumption of power by the Sandinistas in Nicaragua, the nation has begun an English-language international service known as the Voice of Nicaragua. At the time this was written, only two SW stations for domestic reception still operated in the country. One was Radio Sandino, Managua, on 6200 kHz and the other, Radio Zinica, in Bluefields, on 6120 kHz. The former operates continuously, while the latter signs on at 1100 UTC and signs off at 0600 UTC.

El Salvador and Panama had no shortwave outlets active whatsoever at the time this book was written. YSS, Radio Nacional de El Salvador, has been active in the past on both 5980 and 9555 kHz, and these two frequencies are still listed, so there is a possibility they may return. The situation in Panama is much bleaker; no SW outlets have been active in that country for several years, and there is no evidence, despite persistent rumors, that SW activity will resume there anytime in the future.

The Dominican Republic shares the island of Hispaniola with Haiti. It is home to one of the best-known domestic SW broadcasters, HIUA, Radio Clarin, in Santo Domingo, on 11700 kHz. During the late 1970s, this station broadcast an English-language segment known as "This is Santo Domingo," hosted by Rudy Espinal. This quickly became a cult favorite among SWLs, primarily due to Rudy's refreshing and interesting personality. In the 1980s, some American SWLs began producing a program titled "Radio Earth," which was broadcast over Radio Clarin. Rudy Espinal was again the host of this program, but broadcasts from Radio Clarin were eventually terminated as the Radio Earth organization made plans to begin broadcasting from the island of Curacao in the Netherlands Antilles. Today Radio Clarin is but a shadow of its former self; it can be heard until approximately 0500 UTC with all-Spanish programming.

The two most often heard countries from Latin America are Colombia and Venezuela. At night in North America, the 60-meter tropical band seems packed with stations from these two countries until they begin to sign off around 0500 to 0600 UTC. (Some stations operate all night.) Identifying these two countries on 60-meters is simplified because all Colombian stations have frequencies ending in 0, while all Venezuelan stations use frequencies ending in 5. Numerous other Colombian and Venezuelan stations can be found on other frequency bands, but if you're a new SWL the action on 60-meters alone will keep you busy for quite a while. Table 7-1 lists some commonly heard Colombian and Venezuelan stations.

Ecuador is easy to hear via HCJB. More challenging—and, to many SWLs, more interesting—are the numerous domestic SW broadcasters found there. Among the more widely heard are HCFF1, Radio Jesus del Gran Poder, in Quito, on 5050 kHz, HCQR1, Radio Quito, on 4920 kHz, and HCMV5, Radio Popular, in Cuenca, on 4800 kHz.

Brazil breaks the pattern in Latin American broadcasting in several ways. As mentioned before, the prime language here is Portuguese instead of Spanish. And while Brazil has numerous stations on the tropical broadcasting bands, its domestic broadcasters also make wide use of higher-frequency ranges. One of the best bands for reception of Brazilian stations in North America is the 25-meter international broadcasting band. Among the stations that can be heard in the evenings in North America are Radio Globo, in Rio de Janeiro, on 11805 kHz, Radio Bandeirantes, in Sao Paulo, on 11925 kHz, and Radio Nacional Amazonia, in Brasilia, on 11780 kHz. Another well-heard station is Radio Inconfidencia, in

TABLE 7-1 COLOMBIA AND VENEZUELAN STATIONS ON 60-METERS

kHz	Station and location
5095	HJGG, Radio Sutatenza, Bogota, Colombia
4990	YVMQ, Radio Barquisimeto, Barquisimeto, Venezuela
4980	YVOC, Ecos Del Torbes, San Cristobal, Venezuela
4970	YVLK, Radio Rumbos, Villa de Cura, Venezuela
4955	HJCQ, Radio Nacional, Bogota, Colombia
4945	HJDH, Radio Colosal/Caracol, Neiva, Colombia
4905	HJAG, Emisora Atlantico, Barranquilla, Colombia
4900	YVNK, Radio Juventud, Barquisimeto, Venezuela
4880	YVMS, Radio Universo, Barquisimeto, Venezuela
4865	HJLZ, La Voz de Cinaruco, Arauca, Colombia
4855	HJIG, Ondas del Meta, Villavicencio, Colombia
4850	YVKX, Radio Capital, Caracas, Venezuela
4830	YVOB, Radio Tachira, San Cristobal, Venezuela
4780	YVLA, La Voz de Carabobo, Valencia, Venezuela
4755	HJEU, Emisora Nuevo Mundo, Bogota, Colombia

Belo Horizonte, on 15190 kHz. (You'll notice that no call letters were given for these stations. It wasn't an oversight—the information available on them is incomplete and in some cases conflicting.)

Peru is a bit more difficult to hear, but several stations can be heard with a little effort. One that stands out due to its odd frequency is Radio Continente in Juanjui on 8925 kHz; this frequency varies slightly, and the station is not always active. A more "normal" station is OAX4W, Radio America, in Lima, on 6010 kHz; this operates on a 24-hour schedule and can be heard all night if QRM permits. A well-heard 60-meter station is OAX8F, Radio Atlantida, in Iquitos, on 4790 kHz until 0500 UTC sign-off.

Bolivia can be a surprisingly difficult country to hear. Many of its domestic SW stations are low-powered or are covered by QRM from stronger stations in Colombia or Venezuela. Perhaps the best bet is CP92, Radio Panamerica, in La Paz, on 6105 kHz at their 1030 UTC sign-on or in the evenings until their 0400 UTC sign-off.

Reception of Paraguay can be tough even for experienced DXers. However, the government's Radio Nacional, in Asuncion, on 9735 kHz puts a good signal into North America most evenings until its scheduled 0300 UTC sign-off.

The situation is not much better concerning Uruguay. Perhaps the best targets are CXA19, Radio El Espectador, in Montevideo, on 11835 kHz until its 0400 UTC sign-off or CXA3, La Radio, also in Montevideo, on 6035 kHz until 0500 UTC.

Chile is relatively easy via CE1514, Radio Nacional, in Santiago, on 15140 kHz in late afternoons and early evenings; you may also catch some English used here as well. Two private stations which can be heard if QRM from international broadcasters is reduced are CE957, Radio Diego Portales, on 9570 kHz (this frequency varies somewhat) and CE963, Radio Agricultura, on 9630 kHz. Both are in Santiago and operate until 0400 UTC.

Argentina has an international broadcasting service known as Radiodifusion Argentina al Exterior, a holdover from the days of Juan Peron. All domestic SW stations in Argentina are currently government-owned and -operated. The main domestic SW service, known as Radio Nacional, can be heard on 6060 kHz throughout the night until 1230 UTC sign-off.

8
UTILITY STATIONS

As mentioned in the first chapter, utility stations "do work" and are not intended for reception by the general public. That doesn't keep SWLs and DXers from avidly tuning in; that's hardly surprising, since the majority of stations heard on a general-coverage receiver are utility stations. Many utility-station listeners would also contend that many of the most interesting stations you can hear happen to be utilities.

While there is no prohibition against listening to utility stations, there are some restrictions on what you can do with any messages (or "traffic") you hear. Specifically, federal law prohibits divulging or repeating the contents of anything you hear transmitted over a utility station (with the exception of a *marker* transmission, which will be discussed later) to anyone other than the station itself. This means, for example, that *what* was heard over a utility station should not be reported in a SWL club bulletin. You can disclose the time and frequency you heard a utility station, its call sign (and the call signs of any stations it contacted), mode of transmission, and a general idea of what the transmission involved but not its specifics. For example, it would be okay to mention that a utility station you heard was sending telegrams, but it would not be permissible to disclose the contents of any telegram, to whom they were sent, or who sent them. The maximum penalty for violating this federal statute is a $10,000 fine and a year in jail.

Your author knows of no one who has ever been prosecuted for violation of this particular law. In fact, the trend in recent years has been toward a rather flagrant and open disregard for it. For example, virtually all major news organizations listened in on Coast Guard and Navy frequencies during the search for the wreckage of the Space Shuttle *Challenger*. In fact, the first word that the remains of the astronauts had been located came not from NASA but from monitoring of the search ship radio communications. The three major broadcast news networks, along with Cable News Network, did more than merely paraphrase the contents of such transmissions; they rebroadcast them as received. However, the law is still on the books, and efforts have been made to legislate stronger and more encompassing versions.

You'll hear every mode mentioned in Chapter 2 used by utility stations, although FM is mainly restricted to frequencies above 30 MHz, and AM is almost unknown except in the 108- to 136-MHz aeronautical band. The easiest modes for a new SWL to listen to are CW and SSB; most SSB stations operate with USB. RTTY SWLing has recently become extremely popular due to interface devices which allow popular microcomputers to decipher and display RTTY signals. In fact, the majority of fixed stations in operation today use RTTY. Entire books could be—and have been—written on RTTY SWLing. Space considerations will restrict the coverage here to CW, SSB, AM, and FM utility stations. The appendix lists books on RTTY.

Many of these same interfaces can also decipher and display CW transmissions. However, you do not need one of these for CW SWLing if you don't know Morse code. Most CW stations can be identified with the aid of a tape recorder and the Morse code chart in the appendix.

LONGWAVE BEACONS AND OTHER STATIONS

Tune across the range from 200 to 400 kHz at night, and you'll hear stations which seem to do nothing but repeat one to three characters in Morse code continuously—and that Morse code is transmitted in AM using audio tones. You'll also find some stations giving weather forecasts with Morse code in the background. These and similar stations are longwave (LW) beacons. As the name suggests, beacons are used primarily for navigation. At longwave frequencies, loop antennas (discussed in Chapter 4) can be used for direction-finding; by using multiple beacons, ships and aircraft can navigate. Those beacons which include weather information usually do so at 15 and 45 minutes after the hour, although sometimes this is more frequent.

Call signs of beacons don't follow international allocations. Instead, they usually suggest the name of the town or airport at which the beacon is located. Table 8-1 is a sampling of beacons; note the calls. Most suggest the location, but the origin of some are obscure.

Trying to figure out where a beacon is located can be a problem. Beacons normally don't include their location, and beacons change call and frequency, and leave or go on the air with bewildering speed. Fortunately, there are two reference books available from SWL equipment dealers, *The Beacon Guide* by Ken Stryker and *The Radiobeacon Handbook* by Jorgen Trochimczyk. Latest information about new, deleted, or changed beacons can be obtained from SWL club bulletins;

TABLE 8-1 Typical Longwave Beacons

kHz	Call sign and location
194	TUK, Nantucket, Massachusetts
216	CLB, Carolina Beach, North Carolina
263	YGK, Kingston, Ontario, Canada
286	MP, Montauk Point, New York
305	X, Sandusky, Ohio
330	DC, Washington, District of Columbia
344	O, Ottawa, Ontario, Canada
350	ME, Chicago, Illinois (O'Hare Airport)
362	LYL, Lima, Ohio
396	ZBB, Bimini, Bahamas

The Longwave Club of America does a particularly good job of tracking such information.

You'll also notice a few of the stations in Table 8-3 from 415 to 512 kHz, with 500 kHz reserved for emergency traffic. These stations will all be in "true" CW (that is, you'll need your receiver's BFO to copy the signals) and you'll find the communications involve coastal (shore) stations and ships at sea. Coastal stations normally have three-letter call signs, while ship stations normally have calls involving four letters.

It is here that you'll first notice marker transmissions. A marker is the radio equivalent of someone at a microphone saying "Testing, one, two, three." A marker is transmitted by a coastal station repeatedly to allow ship stations to tune it in, to announce that it is on the air and the frequencies it is listening on, and to "hold" a frequency between communications with ships. Markers on CW have the following form:

CQ CQ CQ DE UFL UFL UFL K
VVV VVV VVV DE KFS KFS KFS K

"CQ" is an old radio signal which is a general call to anyone who might be listening; it is an invitation for any listening station to call the transmitting station (in this example, UFL in Vladivostok, USSR). "DE" is French for "from," and is followed by the station's call sign. The "K" is the radiotelegraph equivalent of "over," and signals the end of the transmission. Some markers begin with "VVV." This has no meaning, and simply indicates a test transmission. You might also find "QRA" and "QSX" used in a marker. In such a context, "QRA" asks for the call signs of any stations who might be listening and want to make contact, while "QSX" indicates that the station is listening for replies.

Markers are easy to recognize. The code speed is usually slower than normal CW communications, and the repeated sequence of characters stands out. You'll soon learn to recognize the sound of CQ ("dahdidahdit dahdahdidah"), VVV ("dididah dididah dididah") and DE ("dahdidit dit") in Morse code. With the Morse code table in the appendix and a tape recorder, you can replay the

TABLE 8-3 CW Maritime Station Call Signs and Locations

Call	Location
AME3	Madrid, Spain (Spanish Navy)
CCS	Santiago, Chile (Chilean Navy)
CFH	Halifax, Nova Scotia, Canada
CLA	Havana, Cuba
CTU	Monsanto, Portugal (Portuguese Navy)
CUL	Lisbon, Portugal
C6N	Nassau, Bahamas
DAF	Norddeich, West Germany
DZU	Manila, Philippines
EAD	Aranjuez, Spain
EBC	Cadiz, Spain (Spanish Navy)
FFL	St. Lys, France
FUE	Brest, France (French Navy)
FUF	Fort de France, Martinique (French Navy)
GKE	Portishead, Great Britain
GYC	London, Great Britain (British Navy)
GYQ	Portsmouth, Great Britain (British Navy)
GYU	Gibraltar (British Navy)
HEB	Berne, Switzerland
HKC	Buenaventura, Colombia
HLG	Seoul, South Korea
HPN	Balboa, Panama
HWN	Houilles, France (French Navy)
IAR	Rome, Italy
ICB	Genoa, Italy
KFS	Palo Alto, California
KLB	Seattle, Washington
KLC	Galveston, Texas
LFW	Rogaland, Norway
LSA	Boca, Argentina
LSO	Buenos Aires, Argentina
LZW	Varna, Bulgaria
JCK	Kobe, Japan
JCT	Choshi, Japan
JOS	Nagasaki, Japan
MTI	Plymouth, Great Britain (British Navy)
NAM	Norfolk, Virginia (U.S. Navy)
NMC	San Francisco, California (U.S. Coast Guard)
NMF	Boston, Massachusetts (U.S. Coast Guard)
NMO	Honolulu, Hawaii (U.S. Coast Guard)
NMR	San Juan, Puerto Rico (U.S. Coast Guard)
NPG	San Francisco, California (U.S. Navy)
NPN	Agana, Guam (U.S. Navy)
OMC	Bratislava, Czechoslovakia
OST	Oostende, Belgium
OXZ	Lyngby, Denmark
PCH	Scheveningen, Netherlands
PJC	Curacao, Netherlands Antilles
PPR	Rio de Janeiro, Brazil

TABLE 8-3 CW Maritime Station Call Signs and Locations (Cont.)

Call	Location
PWB	Balem, Brazil
PWZ	Rio de Janeiro, Brazil (Brazilian Navy)
RIH	Khiva, Uzbek SSR, USSR (Soviet Navy)
ROT	Moscow, USSR (Soviet Navy)
SAB	Goteburg, Sweden
SPH	Gydnia, Poland
SVD	Athens, Greece
SVF	Athens, Greece
TIM	Limon, Costa Rica
VHR	Darwin, Australia (Australian Navy)
VRT	Hamilton, Bermuda
UAT	Moscow, USSR
UDH	Riga, Latvian SSR, USSR
UFL	Vladivostok, USSR
UJE	Moscow, USSR (Soviet Navy)
UJY	Kaliningrad, USSR
UMV	Murmansk, USSR
URD	Leningrad, USSR
UXN	Arkhangelsk, USSR
VIP	Perth, Australia
VWB	Bombay, India
WCC	Chatham, Massachusetts
WKM	New Haven, Connecticut
WLO	Mobile, Alabama
WMH	Baltimore, Maryland
WNU	Slidell, Louisiana
WOE	Lantana, Florida
WPA	Port Arthur, Texas
WPD	Tampa, Florida
WSL	Amagansett, New York
XFE	Ensenada, Mexico
XFM	Manzanillo, Mexico
XSG	Shanghai, China
Y5M	Ruegen, East Germany
ZLP	Irirangi, New Zealand
ZSC	Cape Town, South Africa
4XZ	Haifa, Israel
6WW	Dakar, Senegal (French Navy)
6YI	Kingston, Jamaica
8PO	Barbados
9YL	North Post, Trinidad

marker until you have it completely decoded. (You'll also learn the Morse code without meaning to.)

By the way, there are other beacons scattered above and below the standard broadcast band (540 to 1600 kHz) which are similar to those found on LW. Since QRM on these frequencies is low, they can often be heard at surprising distances. For example, your author once heard beacon NB, in North Bay, Ontario, Canada,

Maritime Communications

on 530 kHz on a car radio while driving one night on the New Jersey Turnpike—with an excellent signal! Other well-heard beacons below the broadcast band include UOC in Iowa City, Iowa on 524 kHz and GF in Cleveland, Ohio on 521 kHz. Beacons above the broadcast band propagate very well. RAB, in Rabinal, Guatemala, on 1613 kHz is often heard throughout North America during winter nights. Other beacons that are heard widely include LAG in Lago Agrio, Ecuador on 1665 kHz and MER, Mercaderes, Colombia on 1685 kHz.

MARITIME COMMUNICATIONS

Hundreds of frequencies are used for maritime radio. In fact, more stations are probably active in the maritime service than in any other type of shortwave utility.

Most SSB maritime transmissions will be in USB and will be *duplex*. This means that the ship station and the coastal stations it contacts will be using separate frequencies rather than the same frequency. By international agreement, such frequency pairs have been standardized and assigned reference numbers. As you tune through the maritime SW bands, you'll run across numerous "one-sided" communications; these mean you're listening to one-half of a duplex pair.

There are also some *simplex* frequencies, on which both stations use the same frequency and take turns transmitting and receiving in a normal fashion. Table 8-2 lists some of the more active simplex frequencies. The so-called "call-

TABLE 8-2 Important SSB Marine Radio Frequencies

kHz	Use
2182	International distress and calling
2670	U.S. Coast Guard weather broadcasts
4125	Ship calling and intercoastal waterway traffic
4143.6	Ship and coastal stations
4376	U.S. Coast Guard traffic
4419.4	Coast stations
6218.6	Ship stations
6221.6	Ship and coastal stations
6518.8	U.S. Coast Guard traffic
8257	Ship calling
8291.1	Ship and coastal stations
8294.2	North American inland and intercoastal ships
8718.9	U. S. Coast Guard traffic
8765.4	U.S. Coast Guard traffic
12429.2	U.S. Coast Guard traffic; civilian ships and coastal stations
12435.4	Ship and coastal stations
16523.4	Ship stations
16587.1	U.S. Coast Guard traffic; civilian ships and coastal stations
22124	Ship and coastal stations

ing" frequencies are used to establish contact with another ship or a coastal station; once contact is established, the stations move to another frequency.

Two frequencies that can provide much interesting listening at night if you're within a few hundred miles of a major body of water are 2182 and 2670 kHz. 2182 kHz is the international distress and calling frequency. You won't have to listen long here before hearing a call from a vessel in trouble. Usually, the problem is relatively minor, such as a disabled engine, but sometimes true life-and-death emergencies (such as a sinking ship) can be heard. This frequency is monitored continuously by the U.S. Coast Guard and other services, as well as by other ships at sea. The 2670-kHz channel is used by the U.S. Coast Guard for its marine weather broadcasts. Coast Guard stations along the Atlantic and Pacific coasts transmit these bulletins several times each hour. In between, 2670 kHz is widely used by the Coast Guard for routine communications between its vessels and shore bases. In fact, 2670 kHz is almost never quiet at night; something always seems to be going on there.

The remaining frequencies in Table 8-2 are filled with a variety of merchant vessel and coastal station traffic. Some of the more interesting listening on these channels comes during unusual or difficult weather situations; a ship caught in a squall or hurricane provides dramatic listening.

Most maritime communications are still carried on by CW, however. All maritime bands above 4000 kHz have substantial CW segments, and you'll find most of these humming with activity. The activity you'll hear is very similar to that heard on the longwave marine segment, with standard call signs used for coastal and ship stations along with plenty of marker transmissions. Table 8-3 gives the call signs and locations for some of the more commonly heard CW maritime stations; note that you can hear both the American and Soviet navies.

Many maritime stations will even verify SWL reports. Some have their own printed QSL cards, while others require SWLs to submit a prepared card with their report which the station can then sign and return. Other stations will not verify or acknowledge reception reports under any circumstances, even for marker transmissions. (As you might expect, the Soviet Navy does not acknowledge reception reports under any circumstances.) Addresses for utility stations and details on their verification policies can be found in various SWL publications and club bulletins.

FIXED STATIONS

This is one category of stations where DX opportunities have markedly declined in recent years. Two decades ago, there were numerous fixed stations used for overseas telephone calls. Such stations signed on with a voice marker transmission that went something like "This is a test transmission for circuit adjustment purposes from a station of the American Telephone and Telegraph Company. This station is located near New York City." Such stations often verified reception reports, and QSLs from these outlets are collector items today.

Only a few of these stations still remain, as satellites have taken over the bulk of international telephone traffic. Today, most fixed stations use RTTY. Some fixed stations do use SSB and CW, however, and these stations are listed in Table 8-4.

Among the more fascinating fixed stations are those used for diplomatic communications. Under international law, a nation's embassy or consulate is recognized as the sovereign territory of that nation. Thus, an American embassy in a foreign country is under American "rule," while another nation's embassy in the United States is, in effect, a tiny bit of that nation in the United States. Such embassies and consulates have the right to communicate with their home government using coded messages. Most often wire or satellite links are used, but radio communication may be maintained as well. The United States and other countries have taken advantage of this and have established radio communications facilities at their embassies and consulates. (Some nations have been especially willing to do this, as anyone who has seen the roof of the Soviet embassy in Washington knows.)

Table 8-4 shows some active frequencies and call signs of U.S. embassies in various nations. Often these can be heard running CW markers beginning with "QRA," as in "QRA QRA QRA DE KKN44 KKN44 KKN44 K." One puzzling aspect of these stations are the nations in which they are located, such as Great Britain and Japan. Why are embassy radio stations necessary in such locations? Both nations have abundant, secure satellite and wire links available to the United States, and both are stable allies in which such more reliable communications channels are unlikely to be disrupted by a sudden revolution. It seems likely these stations are actually used for some purpose other than normal diplomatic communications. (One rumor says that some of these stations are actually not con-

TABLE 8-4 Fixed Stations of the World

kHz	Call, location, other data (mode)
4886	KKN44, U.S. Embassy, Monrovia, Liberia (CW)
7719	KNY23, Czechoslovak Embassy, Washington, D.C. (CW)
9040	KNY25, Roumanian Embassy, Washington, D.C. (CW)
9311.5	KEC96, Federal Bureau of Investigation (FBI), New York, N.Y., along with other FBI stations (USB)
10390	FSB57, International Police Organization (Interpol) headquarters, Paris, France, plus several other Interpol stations worldwide (CW)
10463.5	KWL90, U.S. Embassy, Tokyo, Japan (CW)
10493	WGY900, Federal Emergency Management Agency (FEMA), Washington, D.C., along with other FEMA stations nationwide (USB)
12022.5	KKN50, U.S. State Dept., Washington, D.C. (CW)
13815	KRH50, U.S. Embassy, London, Great Britain (CW)
14360	KWS78, U.S. Embassy, Athens, Greece (CW)
20753	HBC88, International Red Cross Headquarters, Geneva, Switzerland, and several other Red Cross stations worldwide (CW, USB, and LSB)

nected with an American embassy at all, but are rather used by other U.S. government agencies as a "cover" for their operations.)

KKN50 is the main U.S. Department of State radio station, and it can be heard running CW markers on numerous frequencies. This station is supposedly used to communicate with U.S. embassies overseas. However, persistent rumors claim that this station is actually involved in some sort of intelligence-gathering activity, with some reports placing actual operation of the station in the hands of the Central Intelligence Agency. While licensed to Washington, actual transmissions take place from a site near Warrenton, Virginia. As will be discussed later, some very interesting signals also originate from the same location.

Foreign embassies in the United States also maintain transmitting facilities. In some cases, the Federal Communications Commission has allocated these stations call signs from the U.S. allocations. Some stations use these, while others do not; under international law, there is no requirement for an embassy to use the call sign supplied by a host government, or even to use any call sign whatsoever. Table 8-4 shows some foreign embassies in Washington that have been active recently in using FCC call signs. Other embassies are certainly active but do not use identifiable call letters. One example is again the Soviet embassy, which has been reportedly supplied the call sign KNY31. You won't find reports of this call being heard in SWL publications or club bulletins. However, the station at the Soviet embassy in Washington is certainly active. Your author has stayed at a hotel adjacent to that embassy and has observed that one large directional transmitting antenna mounted on a rotor changes positions during the day. One position seemed to be that for a path into Central America and the northern portion of South America, while another was ideal for a North Pole path into the USSR; a more puzzling one seemed excellent for a path to the West Coast of the United States.

Consulates might be termed "branch embassies." Usually, these are nothing more than a suite of offices in an ordinary building. Some nations, particularly those in the Soviet bloc, have far more elaborate and secure facilities. For example, the Soviet consulate in San Francisco is every bit as secure and inaccessible as its embassy in Washington—and has an almost equally imposing array of antennas. (A minor mystery a few years ago concerned what was underneath some white plastic "bubbles" erected on the roof of the consulate; these were believed to be antennas of some sort.) If you're in a major city, you might want to check if any consulates are listed in the telephone directory; if so, inspect the buildings where they are visually for antennas. The presence of antennas, particularly any similar to those used by radio amateurs for SW communication, may indicate the presence of transmitting facilities in the consulate. (Such antennas can also be spotted atop some United Nations missions in New York City.)

SWLs who live near embassies and consulates and who are equipped with a portable SW receiver might want to keep a lookout for possible activity from them. If a receiver "overloads" with CW, RTTY, or similar signals near an embassy or consulate, then that embassy or consulate is the likely source of them.

Several U.S. government agencies maintain radio networks as a backup to normal wire and satellite networks. While normally not used, such radio networks are often tested. The FBI uses several frequencies, with 9311.5 kHz being the most commonly reported. The Federal Emergency Management Agency coordinates relief efforts in the event of major disasters (floods, hurricanes, and—yes, indeed—nuclear war). Frequent drills and tests are conducted on 10493 kHz as well as other frequencies.

Two international organizations which use SW radio for communications are the International Police Organization (Interpol) and the International Red Cross. Contrary to popular image, Interpol itself has no agents and makes no arrests; its primary function is as a clearinghouse for information about criminal activity. Much of Interpol traffic is in RTTY, although 10390 kHz is widely used for CW. The International Red Cross uses 20753 kHz for communications related to its humanitarian functions.

You'll also notice what seem to be international broadcasting stations using SSB in the fixed bands. These are *feeder* stations, used by some international broadcasters to relay programs from their studios to transmitting sites. These stations are gradually being phased out as satellites link the sites.

TIME AND FREQUENCY STATIONS

WWV, WWVH, and CHU are not the only standard time and frequency stations you can hear. Table 8-5 lists some of those that have been heard in North America. Naturally, European stations are better heard along the East Coast, while the Asian stations are better heard along the West Coast. Like WWV/WWVH and CHU, other time and frequency stations transmit time pulses each second; some also include propagation forecasts and voice announcements as well. Many of these stations also verify receptions reports; Figure 8-1 shows a QSL card received from OLB5 in Czechoslovakia for reception on 3170 kHz.

While WWV/WWVH and CHU identify by voice at regular intervals, such is not the case with other time and frequency stations. Some, like Y3S on 4525 kHz, never identify; you'll only hear pulses each second. Other stations, such as Soviet outlets, identify only in Morse code. And station VNG in Australia identifies in voice, but only on the quarter hour. Nor do all of these stations operate continuously. The latest operating schedules can be found in a current edition of *The World Radio TV Handbook*.

There are additional time and frequency stations throughout the world; some of these operate for as little as five minutes at a time. *The World Radio TV Handbook* contains information on these other stations.

TABLE 8-5 Time and Frequency Stations of the World

kHz	Call and location
2500	JJY, Tokyo, Japan
2500	WWV, Fort Collins, Colorado
2500	ZUO, Olifantsfontein, South Africa
3170	OLB5, Podebrady, Czechoslovakia
3330	CHU, Ottawa, Ontario, Canada
3810	HD210A, Guayaquil, Ecuador
4500	VNG, Lyndhurst, Australia
4525	Y3S, Nauen, East Germany
4996	RWN, Moscow, USSR
5000	IAM, Rome, Italy
5000	IBF, Turin, Italy
5000	JJY, Tokyo, Japan
5000	LOL, Buenos Aires, Argentina
5000	WWV, Fort Collins, Colorado
5000	WWVH, Kauai, Hawaii
6100	YVTO, Caracas, Venezuela
7335	CHU, Ottawa, Ontario, Canada
7500	VNG, Lyndhurst, Australia
7600	HD20A, Guayaquil, Ecuador
8000	JJY, Tokyo, Japan
9996	RWM, Moscow, USSR
10000	BPM, Xian, China
10000	JJY, Tokyo, Japan
10000	LOL, Buenos Aires, Argentina
10000	RCH, Tashkent, Uzbek SSR, USSR
10000	WWV, Fort Collins, Colorado
10000	WWVH, Kauai, Hawaii
10004	RID, Irkutsk, USSR
12000	VNG, Lyndhurst, Australia
14670	CHU, Ottawa, Ontario, Canada
14996	RWM, Moscow, USSR
15000	BPM, Xian, China
15000	JJY, Tokyo, Japan
15000	LOL, Buenos Aires, Argentina
15000	WWV, Fort Collins, Colorado
15004	RID, Irkutsk, USSR
20000	WWV, Fort Collins, Colorado

AERONAUTICAL COMMUNICATIONS

USB is extensively used for aeronautical work between aircraft flying international routes and airports. And language won't prove a major obstacle, since English is the language of international aviation. The only other language you're likely to hear used to any significant extent is French (mainly for flights into and out of Africa) and some scattered Russian (for flights to and from the Soviet Union).

Aeronautical Communications 163

Figure 8.1 QSL card from standard time and frequency station OLB5 on 3170 kHz.

"VOLMET" transmissions are aviation weather broadcasts giving information on weather conditions enroute and at various destinations. Table 8-6 gives the frequencies and originating airports for selected VOLMET transmissions. In most cases, there is more than one VOLMET on a frequency. In such cases, the various stations take turns transmitting. Identification will usually be "This is New York Radio," "This is Shannon Aeradio," or a similar form. Note that the various frequencies are regional and serve flights flying various routes. There are VOLMET frequencies beyond those in Table 8-6; however, those given are among the easiest to receive.

Several frequencies have been set aside for SSB communications by flights along various international routes and airports. Transmissions on these frequencies are essential communications for the safety of the flight, such as arrival times, fuel consumption, weather information, and the like. These frequencies are listed in Table 8-7. Several other frequencies are reserved for long-distance operations control (LDOC) use. These are noncritical, "internal" communications between an airline and its aircraft aloft. Certain frequencies are particularly used by certain airlines. Table 8-8 lists some frequencies and the major airlines that use them.

Some of the most interesting aeronautical listening involves the various branches of the United States military. Table 8-9 lists some of the more widely used frequencies and the various services which use them; all transmissions will be in USB.

One thing you'll quickly note is how fond the various services, particularly

TABLE 8-6 VOLMET Frequencies and Locations

kHz	Locations
2863	Auckland, New Zealand; Hong Kong; Honolulu, Hawaii; Tokyo, Japan
2881	Brasilia, Brazil; Buenos Aires, Argentina; Lima, Peru
2998	Prague, Czechoslovakia; Tel Aviv, Israel
3413	Shannon, Ireland
3485	Gander, Newfoundland, Canada; Long Island, New York
5499	Antanarivo, Madagascar; Brazzaville, Congo; Johannesburg, South Africa; Nairobi, Kenya
5561	Bahrain; Beirut, Lebanon; Cairo, Egypt; Istanbul, Turkey; Tehran, Iran
5601	Brasilia, Brazil; Buenos Aires, Argentina; Lima, Peru
5640	Shannon, Ireland
6580	Prague, Czechoslovakia; Tel Aviv, Israel
6604	Gander, Newfoundland, Canada; Long Island, New York
6676	Bangkok, Thailand; Bombay, India; Calcutta, India; Karachi, Pakistan; Singapore; Sydney, Australia
6679	Auckland, New Zealand; Hong Kong; Honolulu, Hawaii; Tokyo, Japan
8828	Auckland, New Zealand; Hong Kong; Honolulu, Hawaii; Tokyo, Japan
8957	Shannon, Ireland
10051	Gander, Newfoundland, Canada; Long Island, New York
10057	Antanarivo, Madagascar; Brazzaville, Congo; Johannesburg, South Africa; Nairobi, Kenya
10087	Brasilia, Brazil; Buenos Aires, Argentina; Lima, Peru
11378	Prague, Czechoslovakia; Tel Aviv, Israel
11387	Bangkok, Thailand; Bombay, India; Calcutta, India; Karachi, Pakistan; Singapore; Sydney, Australia
13261	Antanarivo, Madagascar; Brazzaville, Congo; Johannesburg, South Africa; Nairobi, Kenya
13264	Shannon, Ireland
13270	Gander, Newfoundland, Canada; Long Island, New York
13279	Brasilia, Brazil; Buenos Aires, Argentina; Lima, Peru
13282	Auckland, New Zealand; Hong Kong; Honolulu, Hawaii; Tokyo, Japan

the Air Force, are of so-called "tactical" call signs. You'll hear stations identifying themselves as "Elastic," "Classical," "Morphine," "Rim Control," and similar fanciful, meaningless terms. Such call signs change on a daily basis. If they leave you confused, that's the entire point; any potential enemies are supposedly equally confused and are unable to identify the various stations or "enter" the communications network without being detected.

Yet some "normal" identifications are used, and it is usually possible to at least determine which armed service you are listening to. The general Air Force identification format for its aircraft consists of a letter from the phonetic alphabet, such as "Delta," followed by two or three digits. Air Force bases and ground stations often refer to themselves simply by their location or name, such as "Andrews" or "Vandenburg." Navy stations follow a similar format, although ground station names are usually preceded by the word "Raspberry."

Aeronautical Communications

TABLE 8-7 International Civilian Aviation Frequencies

kHz	Use
4675	North Atlantic flights
5493	Central and Southern Africa flights
5526	South American flights
5550	Caribbean flights
5589	International air-to-ground frequency
5616	North Atlantic flights
5628	North Pacific flights
5643	South Pacific flights
5667	Middle East flights
6532	West Pacific flights
6535	South Atlantic flights
6556	South Asia flights
6571	Central Asia flights
6598	East European flights
6661	North Pacific flights
8846	West Caribbean flights; South Pacific flights
8855	South American flights
8879	Indian Ocean flights
8891	North Atlantic flights
8903	Central Africa flights
8918	Middle East flights
8942	Southeast Asia flights
11300	North Africa and Middle East flights
11309	North Atlantic flights
11387	Caribbean flights
11396	Central American and Caribbean flights
13288	North Africa flights
13291	North Atlantic flights
13294	African flights
17955	South Atlantic flights
17961	Indian Ocean flights

The Civil Air Patrol (CAP) is a civilian organization sponsored by the Air Force, and is often active for search and rescue missions and drills. Stations in CAP are assigned call signs from normal FCC allocations consisting of three letters followed by three digits. However, on-air activity is usually identified by a tactical call sign consisting of a combination of four letters and digits.

The Tactical Air Command (TAC) is responsible for defense of North America against attack from enemy aircraft. The various TAC ground stations identify themselves by the word "Raymond" followed by one or two digits.

Those frequencies identified in Table 8-9 as belonging to the Strategic Air Command (SAC) will be of particular interest. SAC is responsible for nuclear missiles and bombers, and all of its frequencies in Table 8-9 are part of its "Giant Talk" communications network. Don't expect to understand any of the messages

TABLE 8-8 In-Flight LDOC Channels

kHz	User(s)
4654	Swissair
5532	KLM
6637	Air France, El Al, JAL, Lufthansa, Quantas
8921	British Airways
8924	Aeroflot, El Al, KLM, Varig
10030	Aeroflot
10072	British Airways
10075	Middle Eastern Airlines
11215	Eastern Airlines
11351	Air France

you might hear on a SAC frequency; as you would expect, all are carefully encoded. Often you may hear a transmission on a SAC frequency begin with the word "Skyking." This is generally believed to be a general message to all SAC nuclear forces. These messages are generally quite short, and consist of a few numbers and letters from the phonetic alphabet. What they refer to, and what they mean, are unknown and beyond the ability of a casual listener (and, it is to be hoped, any hostile or potentially hostile nation) to determine.

Listening to military communications has developed into something of a specialty within the SWLing hobby. The best way to keep up to date on happenings in this field is to join a SWL club with good utility coverage.

TABLE 8-9 U.S. Military Aviation Frequencies

kHz	Use
4464.5	Civil Air Patrol
4495	U.S. Air Force
4582	Civil Air Patrol
4700	U.S. Navy
4725	U.S. Air Force/Strategic Air Command
4732	Air Force One
4737	U.S. Navy
5020	U.S. Air Force
5297	North American Aerospace Defense Command (NORAD)
5500	Civil Air Patrol
5692	U.S. Coast Guard; U.S. Navy
5700	U.S. Air Force/Strategic Air Command
6670	U.S. Air Force
6697	U.S. Navy
6701	U.S. Navy
6720	U.S. Navy
6730	U.S. Air Force/Tactical Air Command; Air Force One
6750	U.S. Air Force
6756	Air Force One

Aeronautical Communications

TABLE 8-9 U.S. Military Aviation Frequencies (Cont.)

kHz	Use
6761	U.S. Air Force/Strategic Air Command
6833	U.S. Navy
8101	U.S. Air Force/Strategic Air Command
8972	U.S. Navy
9002	U.S. Navy
9014	U.S. Air Force/Tactical Air Command
9018	Air Force One
9023	U.S. Air Force/Strategic Air Command
9027	U.S. Air Force/Strategic Air Command
9036	U.S. Navy
9057	U.S. Air Force
11118	Air Force One
11191	U.S. Navy
11243	U.S. Air Force/Strategic Air Command
11258	U.S. Navy
11407	U.S. Air Force
11494	U.S. Air Force
13201	U.S. Air Force; Air Force One
13204	Air Force One
13215	U.S. Air Force; Air Force One
13231	U.S. Navy
13241	U.S. Air Force/Strategic Air Command
13251	U.S. Navy
13440	Air Force One
13907	U.S. Air Force
14913	Air Force One
14955	U.S. Air Force/Strategic Air Command
15041	U.S. Air Force/Strategic Air Command
15091	U.S. Air Force/Tactical Air Command
17975	U.S. Air Force/Strategic Air Command
17982	U.S. Navy
17985	U.S. Navy
18009	U.S. Navy
18027	Air Force One
18594	U.S. Air Force
20631	U.S. Air Force/Strategic Air Command
23337	U.S. Air Force/Strategic Air Command

9

OTHER RADIO ACTIVITIES

As noted earlier, the hobby of "shortwave listening" involves more than just listening to stations transmitting from 1600 to 30000 kHz. In this chapter, we'll look at some "nonshortwave" SWL activities, some of which involve *transmitting* as well as receiving.

BROADCAST BAND DXING

Many SWLs (such as your author) first experience reception of distant radio signals on the AM broadcast band (abbreviated BCB), which ranges from 540 to 1600 kHz. Some SWLs stick with it and make DXing the BCB their prime listening interest. A few top BCB DXers have managed to hear over 100 different countries entirely between 540 to 1600 kHz.

It's easy to get started in BCB DXing. Tune across the BCB any time between your local sunset and sunrise, and you'll hear several stations located hundreds or even thousands of miles away. Many of these stations verify reception reports, and with a little effort and time you can probably collect QSLs from stations in 25 different states as well as from neighboring countries such as Canada, Cuba, and Mexico. In fact, any ordinary radio tuning the BCB should be adequate to let you hear half of the states, several Canadian provinces, and perhaps ten different countries.

But once you reach those totals, you'll find yourself hitting a wall. You'll notice that the same stations tend to be heard on the same frequencies each night, with only minor variations in what can be heard from day to day. To increase your totals of states and countries heard, you'll find that specialized equipment, anten-

Broadcast Band Dxing

nas, and knowledge are needed, and that geography plays a crucial role in the results you can achieve. A BCB DXer located near the shore in a lightly populated area of the East Coast will *always* be able to hear more countries than a DXer with similar skills and the same equipment located in the Midwest. In fact, probably the best location for a BCB DXer in terms of reception possibilities would be on an island in the middle of an ocean far from major land masses, such as Hawaii or Tahiti.

One reason BCB DX is so challenging was discussed in Chapter 5. The ionosphere always takes a bigger "bite." out of a signal at BCB frequencies than it does for signals at higher frequencies. Multihop reception is common on SW frequencies, but such reception involving more than a couple of "bounces" off the layers of the ionosphere are rare on the BCB. When it does happen, signals are subject to long periods of fading; you may find that out of a five-minute listening period, only one minute of audio can be heard. Simply put, the ionosphere is usually kind to signals at SW frequencies but is murder on BCB frequencies.

Noise is another problem. Lightning from thunderstorms produces radio noise (or "QRN"—remember?). It so happens that this noise, which is a radio signal, propagates rather well at BCB frequencies. It's not unusual for QRN from thunderstorms hundreds of miles distant to be heard—a listener in Chicago in the dead of winter can experience QRN from thunderstorms in the Gulf of Mexico. DXing the BCB often means a case of "ringing" ears from QRN; a good noise blanker circuit in your receiver is a must for serious work.

The most formidable challenge, however, is the sheer number of stations found on the BCB. In North and South America, the BCB is divided into 107 "channels" spaced every 10 kHz beginning at 540 kHz and extending to 1600 kHz. In the United States alone, over 5000 stations are crammed into this range! Many of these stations only operate in the daytime; most others reduce power or use directional antennas at night to avoid causing QRM to other stations on the frequency. Despite these measures, you'll find most channels will have several stations operating on them at night. Some form of directional antenna, such as a rotatable indoor loop, is a major help in sorting out these stations.

Not all BCB stations operate on 10-kHz spacing, however. Some stations in the Caribbean and South America operate on what are known as "split" frequencies between the normal 10-kHz spacing, such as 655 and 1555 kHz. Sometimes this is a deliberate choice by the station to avoid QRM, while in other cases it is the result of improper transmitter operation. Outside North and South America, BCB channels are spaced every 9 kHz rather than 10 kHz. This means that distant BCB stations in Europe, Africa, and Australia can sometimes "sneak through" between 10-kHz channels in North America. Since such stations may operate only one kHz or two from a loud North American BCB station, you'll need a receiver with superb selectivity to have any realistic chance at these stations.

In BCB DX terminology, a "domestic" station is considered to be any AM station in Canada as well as the United States. Domestic BCB stations fall into three categories: *clear channel*, *regional*, and *local* stations. This latter classification is sometimes referred to as "graveyard" frequencies. These are 1230, 1240, 1340, 1400, 1450, and 1490 kHz; stations on these channels in the United States are authorized to use 1 KW of transmitter power and nondirectional antennas around the clock. Each channel has close to 200 stations operating in the United States, with the result that QRM is tremendous. Spend just a few minutes listening on one of these channels where you don't have a local station; it will probably sound something like the "rumble" in a concert hall or theater before the performance begins. DXing these frequencies is a test of your endurance; you may spend hours on a frequency waiting for one station to surface above the jumble long enough for it to be positively identified.

The term "clear channel" is misleading these days. In the past, it was used to denote a station which was the only one on that frequency which could be heard throughout most of the United States and Canada at night. Such stations operated with 50 KW of power and nondirectional antennas. Decades ago, ordinary listeners throughout the country listened to such stations as WLW in Cincinnati or WSM in Nashville on a regular basis in much the same manner that today's viewers watch television "superstations" such as WTBS in Atlanta or WOR in New York. Over the years, the various clear-channel monopolies were broken up, primarily by placing additional stations on each "clear" channel and mandating the use of directional antennas to prevent interference between stations. An excellent example of this arrangement is found on 1110 kHz, where WBT in North Carolina and KFAB in Nebraska both operate with 50 KW. However, both use highly directional antenna systems which concentrate most of their power in a north-south pattern and radiate very little energy to the east or west. The effect of this is that each station's "effective" power is much greater to the north or south, in excess of 50 KW, but only a few hundred watts to the east or west. Thus, the two stations can operate without interference to each other in their primary target audience areas.

The end of the clear-channel concept began in the early 1980s. For years, many stations had been authorized to operate on clear channels during daytime; at sunset, they had to leave the air. Beginning around 1980, several of these stations were authorized to continue operations at night with directional antennas and, in many cases, reduced power. In addition, new stations were allowed to begin operation on some clear channels. The result is that much of the former wide coverage areas of clear-channel stations have been eroded, and the process is still continuing. It is still possible to hear many "clears" widely, however; Table 9-1 lists some targets you can try for.

Regional stations are in the middle ground between clear channels and graveyards. These stations operate with powers of 10 KW or less at night (some use powers as high as 50 KW during the daytime) with directional antennas. Many regional stations use different antenna patterns for day and night operation.

The trend in broadcasting has been toward continuous operation for those stations permitted to do so, although a few regional stations sign off around their

TABLE 9-1 Domestic "Clear Channel" DX Targets

kHz	Call	Location
540	CBK	Watrous, Saskatchewan, Canada
540	CBT	Grand Falls, Newfoundland, Canada
540	CJSB	Ottawa, Ontario, Canada
630	CHED	Edmonton, Alberta, Canada
640	KFI	Los Angeles, California
650	WSM	Nashville, Tennessee
660	KTNN	Window Rock, Arizona
660	WNBC	New York, New York
670	KBOI	Boise, Idaho
670	WMAQ	Chicago, Illinois
680	CFTR	Toronto, Ontario, Canada
680	KNBR	San Francisco, California
680	WCNN	Atlanta, Georgia
680	WPTF	Raleigh, North Carolina
680	WRKO	Boston, Massachusetts
690	CBF	Montreal, Quebec, Canada
690	CBU	Vancouver, British Columbia, Canada
700	WLW	Cincinnati, Ohio
710	KIRO	Seattle, Washington
710	KMPC	Los Angeles, California
710	WGBS	Miami, Florida
710	WOR	New York, New York
720	KDWN	Las Vegas, Nevada
720	WGN	Chicago, Illinois
730	CKAC	Montreal, Quebec, Canada
730	CKLG	Vancouver, British Columbia, Canada
740	CBL	Toronto, Ontario, Canada
740	CBX	Edmonton, Alberta, Canada
740	KCBS	San Francisco, California
740	KTRH	Houston, Texas
750	KFQD	Anchorage, Alaska
750	WSB	Atlanta, Georgia
760	KFMB	San Diego, California
760	WJR	Detroit, Michigan
770	KKMI	Seattle, Washington
770	KOB	Albuquerque, New Mexico
770	WABC	New York, New York
780	KROW	Reno, Nevada
780	WBBM	Chicago, Illinois
800	CKLW	Windsor, Ontario, Canada
810	KGO	San Francisco, California
810	WGY	Schenectady, New York
820	KCBF	Fairbanks, Alaska
820	WBAP	Fort Worth, Texas
830	WCCO	Minneapolis, Minnesota
840	WHAS	Louisville, Kentucky
850	KOA	Denver, Colorado
850	WHDH	Boston, Massachusetts
860	CJBC	Toronto, Ontario, Canada

TABLE 9-1 Domestic "Clear Channel" DX Targets (Cont.)

kHz	Call	Location
870	WWL	New Orleans, Louisiana
880	KRVN	Lexington, Nebraska
880	WCBS	New York, New York
890	KDXU	St. George, Utah
890	WLS	Chicago, Illinois
900	CHML	Hamilton, Ontario, Canada
1000	KOMO	Seattle, Washington
1000	WCFL	Chicago, Illinois
1010	CBR	Calgary, Alberta, Canada
1010	CFRB	Toronto, Ontario, Canada
1010	WINS	New York, New York
1020	KBCQ	Roswell, New Mexico
1020	KDKA	Pittsburgh, Pennsylvania
1020	KTNQ	Los Angeles, California
1030	KTWO	Casper, Wyoming
1030	WBZ	Boston, Massachusetts
1040	WHO	Des Moines, Iowa
1050	CHUM	Toronto, Ontario, Canada
1050	WHN	New York, New York
1060	CFCN	Calgary, Alberta, Canada
1060	KYW	Philadelphia, Pennsylvania
1070	CBA	Moncton, New Brunswick, Canada
1070	KNX	Los Angeles, California
1070	WIBC	Indianapolis, Indiana
1080	KRLD	Dallas, Texas
1080	WTIC	Hartford, Connecticut
1090	KAAY	Little Rock, Arkansas
1090	KING	Seattle, Washington
1090	WBAL	Baltimore, Maryland
1100	KFAX	San Francisco, California
1100	WWWE	Cleveland, Ohio
1110	KFAB	Omaha, Nebraska
1110	KRLA	Pasadena, California
1110	WBT	Charlotte, North Carolina
1120	KMOX	St. Louis, Missouri
1120	KPNW	Eugene, Oregon
1130	CKWX	Vancouver, British Columbia, Canada
1130	KSDO	San Diego, California
1130	KWKH	Shreveport, Louisiana
1130	WCXI	Detroit, Michigan
1130	WDGY	Minneapolis, Minnesota
1130	WNEW	New York, New York
1140	CKXL	Calgary, Alberta, Canada
1140	KRAK	Sacramento, California
1140	WRVA	Richmond, Virginia
1150	CKOC	Hamilton, Ontario, Canada
1160	KSL	Salt Lake City, Utah
1170	WWVA	Wheeling, West Virginia
1180	WHAM	Rochester, New York

Broadcast Band Dxing

TABLE 9-1 Domestic "Clear Channel" DX Targets (Cont.)

kHz	Call	Location
1190	KEX	Portland, Oregon
1190	WOWO	Fort Wayne, Indiana
1200	WAGE	Leesburg, Virginia
1200	WOAI	San Antonio, Texas
1210	WCAU	Philadelphia, Pennsylvania
1220	CKDA	Victoria, British Columbia, Canada
1220	WGAR	Cleveland, Ohio
1500	KSTP	St. Paul, Minnesota
1500	WTOP	Washington, District of Columbia
1510	KGA	Spokane, Washington
1510	WLAC	Nashville, Tennessee
1510	WMRE	Boston, Massachusetts
1520	KOMA	Oklahoma City, Oklahoma
1520	WKBW	Buffalo, New York
1530	KFBK	Sacramento, California
1530	WCKY	Cincinnati, Ohio
1540	KXEL	Waterloo, Iowa
1550	CBE	Windsor, Ontario, Canada
1560	WQXR	New York, New York
1570	CKLM	Montreal, Quebec, Canada
1580	KDAY	Los Angeles, California

local midnight and resume operation near sunrise. Many domestic stations are still permitted only daytime operation, although some are permitted operation one or two hours before their local sunrise and after their local sunset at greatly reduced power levels (sometimes below 100 watts).

These patterns of sunrise/sunset changes—power levels, antenna patterns, stations returning to or leaving the air—make sunrise and sunset favorite listening times for domestic DX. Listening at such times requires one to be alert. For example, a station may sign off or abruptly change its power or antenna pattern, leaving a second station "in the clear." This situation may last only a few minutes until the second station likewise adjusts its power or antenna pattern or signs off, leaving a third station present on the frequency. Those few minutes could well be your only chance to hear the second station.

In years past, many stations which broadcast on a continuous schedule would go silent on Monday mornings from their local midnight until their local sunrise. This opened up their channels to enable other stations to be heard. In addition, daytime-only stations are permitted to conduct equipment tests from their local midnight to sunrise; this is known as the "experimental period." This provided a boon to BCB DXers, as stations which were normally impossible to hear could often be heard through the country conducting equipment tests on Monday mornings. Sadly, this is now largely a thing of the past. Many stations today have greatly cut back on the equipment tests they perform and operate without interruption. In addition, the greater number of stations operating on the BCB means that a frequency is not much clearer of interference if only one or two stations are off the air. As a result, Monday mornings now sound much like any other morning of the week on the BCB. However, your best chances of catching a station off the air or an equipment test still remains on Monday mornings.

While the chances for domestic BCB DX have been declining, there are still opportunities to hear foreign stations. Since so many foreign BCB stations operate on split frequencies, receiver selectivity is important; a receiver equipped with a narrow bandwidth mechanical or crystal filter will be an enormous help. In many situations, exalted carrier SSB reception (described in Chapter 3) must be used. Some form of directional antenna, such as a loop, is also important. Most loop designs for sale to BCB DXers (or constructed by DXers) include some amplification to boost the signal before it goes to the receiver, and this is a great help in chasing weak foreign signals. Figure 9-1 shows a popular indoor loop antenna for BCB DX.

Not all foreign BCB DX is a major challenge, however. Quite a few foreign stations can be heard on average communications receivers using ordinary longwire antennas. Table 9-2 lists some of these. As you might expect, persons located in the southern half of the United States will find reception of most of these stations easier, with those DXers located in coastal areas also having an advantage. Not all of these can be heard throughout the United States and Canada, but they do give a good idea of the types of stations that can be heard.

More difficult are stations such as those in Table 9-3. Your chances of hearing any of these is as much a product of where you are as they are of your skills and receiving equipment. Listeners along the East Coast have good opportunities for reception from Europe, Africa, and some of the Middle East but very poor chances for Asia and the Pacific. Listeners along the West Coast have good possibilities for Pacific, Asian, and Australian reception but poor odds for Europe, Africa, and the Middle East. Those BCB DXers in the middle of North America theoretically have a chance to hear all such areas, but their odds of hearing *any* of them are poor.

Some European and African stations operate all night, and thus you can try for them from around your local sunset until sunrise occurs at the transmitter site (usually around 0530 to 0700 UTC). Other stations sign off around 0000 UTC and resume operations in the 0530 to 0630 UTC period. By contrast, DXing Pacific, Asian, and Australian stations is very much an early morning affair. The first stations from such areas do not begin to fade in until around 0800 UTC, and a possible propagation path to many areas does not exist prior to approximately 1000 UTC.

Foreign BCB DX is primarily a fall and winter activity. The lower ionospheric absorption, coupled with longer periods of darkness, means that many

Broadcast Band Dxing

Figure 9.1 Radio West manufactures this compact loop antenna specifically for BCB DX reception.

stations are only realistically possible then. QRN from thunderstorm activity is also reduced compared to the spring and summer months. Late September to early November often has conditions for excellent European, African, and Middle East reception, while March and April are often the best months for Pacific, Asian, and Australian reception. Those interested in domestic DX can chase it all year, since changing patterns of sunrise and sunset allow different targets to be heard, and the possibility of a station's being off the air or conducting an equipment test exists any Monday morning of the year.

Recent years have not been kind to the hobby of BCB DXing, particularly for those chasing domestic stations. The increased numbers of AM stations, and resulting QRM levels, have caused several BCB DXers to abandon the hobby altogether. Some of them have been among those recognized as the most skilled and experienced BCB DXers; apparently, they have literally run out of new countries or stations to hear with any reasonable amount of effort. Clubs specializing in BCB DX give some evidence of this trend, as material describing the "good old days" of BCB DXing several decades ago appears in their pages with increasing regularity. However, there is a possibility that the days of "wide open" frequencies with few stations could be briefly repeated. The Federal Communications Commission has announced its intentions to expand the BCB by an additional 100 kHz, to 1700 kHz, sometime within the 1990s. This expansion awaits

TABLE 9-2 BCB DX Targets in the Americas

kHz	Call	Station name (if any) and location
540	XEWA	La Voz de America Latina, San Luis Potosi, Mexico
555	ZIZ	Basseterre, St Kitts and Nevis
590	CMCA	Radio Rebelde, La Julia, Cuba
600	CMKA	Radio Rebelde, Urbano Noris, Cuba
640	CMBB	Radio Progreso, Guanabacoa, Cuba
655	YSS	Radio El Salvador, San Salvador, El Salvador
675	HCVP2	Sistema de Emisora Atalaya, Guayaquil, Ecuador
690	CMEC	Radio Rebelde, Santa Clara, Cuba
695	HCRS2	Radio Sucre, Guayaquil, Ecuador
705	——	Radio St. Vincent, Kingstown, St. Vincent
730	XEX	Radio Festival, Mexico City, Mexico
735	HCGCI	Radio Melodia, Quito, Ecuador
765	YSKL	Radio Universal, Usulutan, El Salvador
800	PJB	Trans World Radio, Bonaire, Netherlands Antilles
825	——	Radio Paradise, Basseterre, St. Kitts
830	——	Radio Belize, Belize City, Belize
854	OAX4	Radio Nacional del Peru, Lima, Peru
885	ZJB	Radio Montserrat, Plymouth, Montserrat
895	HJPM	Radio Galeon, Santa Maria, Colombia
900	XEW	La Voz de America Latina, Mexico City, Mexico
905	YSQJ	Radiopolis, San Salvador, El Salvador
1015	YSC	Radio Internacional, San Salvador, El. Salvador
1025	HOU	La Voz del Baru, David, Panama
1145	HRQN	La Voz del Atlantico, Puerto Cortes, Honduras
1165	——	Caribbean Radio Lighthouse, St. Johns, Antigua
1255	HCEM1	Ondas Carchenses, Tulcan, Ecuador
1505	——	Radio Anguilla, The Valley, Anguilla
1545	HCDP1	Radio Caracol, Quito, Ecuador
1555	——	Radio Cayman, Gun Bluff, Grand Cayman Island
1570	XERF	"Love 16," Ciudad Acuna, Mexico
1610	——	Caribbean Beacon, The Valley, Anguilla

formal agreements and arrangements with neighboring countries, but already such firms as Motorola are testing antennas and transmitters for this new range. The current plans call for stations in the regional and local categories to fill the new channels beginning at 1610 kHz. Even a modestly powered station on such frequencies could be heard coast to coast if QRM is low, as it will be in the beginning. Alert, active BCB DXers during the first few months and years when the new range is opened will have the chance to make some "once in a lifetime" receptions which will become impossible as more stations crowd the new segment. It might be worthwhile to hone your BCB DXing skills and equipment in preparation for this opportunity!

FM and TV DXing

TABLE 9-3 Transoceanic BCB DX Targets

kHz	Station and location
650	KORL, Honolulu, Hawaii
675	France-Inter, Marseille, France
765	French Language Network, Sottens, Switzerland
774	NHK Second Program, Akita, Japan
846	Radio Two, Rome, Italy
855	Home Service Second Program, Bucharest, Romania
870	KAIM, Honolulu, Hawaii
945	France-Culture, Toulouse, France
1017	TRT-1, Istanbul, Turkey
1017	Sudwestfunk, Rheinsender, West Germany
1035	Chinese First Program, various locations in China
1044	Arabic Home Service, Sebaa-Aioun, Morocco
1215	Radio Three, various locations in Great Britain
1296	BBC External Services, Orfordness, Great Britain
1305	Chinese Second Program, various locations in China
1395	Radio Tirana external services, Lushnje, Albania
1467	Radio Monte Carlo, Monte Carlo, Monaco
1548	4QD, Queensland, Australia
1593	Westdeutscher Rundfunk, Langenberg, West Germany

FM AND TV DXING

Many of those who have given up on BCB DXing have now turned their attention to chasing distant stations on the FM broadcast band and TV channels, particularly channels 2 through 6. Unlike BCB DXing, success in FM and TV DXing depends more upon being on the right frequency at the right time than it does upon receiving equipment or even skill.

Your location does have an impact on the results you can expect, however. DXers in a metropolitan area with numerous FM and TV stations will be at a severe disadvantage, since these strong local stations will block DX signals on the same frequency. Those in more suburban or rural areas will find the DX prospects much better. In such areas, a portable FM radio or television set can provide plenty of DX, even if a simple indoor antenna is used.

One propagational method for FM and TV signals is sporadic-E, which was discussed in Chapter 5. While it can happen at any time of the year, it is most common from late May until early August with another period often possible from late December to early January. (Actually, this second period is the result of sporadic-E clouds from south of the equator drifting northward.) Sporadic-E is a very democratic mode of propagation; large, extensive outdoor antenna arrays are often less effective than simple indoor "telescoping" or "rabbit ears" antennas.

And the signal strengths delivered via "E skip" are so strong that simple FM radios and TV sets do fine.

The MUF of sporadic-E propagation can vary greatly within the space of a few minutes. The effects of sporadic-E are first, and most often, noticed on the citizen's band and 10-meter amateur radio band frequencies, located approximately from 26000 to 29700 kHz. When 10-meters (28000 to 29700 kHz) is filled with signals from 600 to 1500 miles distant, especially during the months when sporadic-E is most likely, it's a good sign that a sporadic-E opening is underway. The MUF the opening eventually reaches can be determined only by listening (or viewing, as the case may be). In the Americas, TV channel 2 begins just 54 MHz and TV channel 6 ends just below 88 MHz; the FM broadcast band extends from 88 to 108 MHz. Thus, the effects of a sporadic-E opening will first be observed on channel 2, and the rise of the MUF can be tracked as channels 3, 4, 5, and 6 "open" for DX reception. When sporadic-E propagation is observed on channels 5 and 6, it's a signal that the MUF may reach into the FM broadcast band. One of the fascinating aspects about sporadic-E is the way the MUF frequently hits the middle of the band, such as 96 MHz. The result in such a case would be a band full of rare DX signals from hundreds or thousands of miles away below 96 MHz and only local FM stations above that. Not every sporadic-E opening reaches the FM band or upper TV channels, however. Most openings will only reach channels 2, 3, or 4. Thus, you'll probably "see" more TV DX than hear FM DX, unless you live in an area where you have one or more stations operating locally on channels 2, 3, and 4.

The first sign of a sporadic-E opening is usually "rolling" black bars across the picture received from a local station on channel 2, 3, or 4. These bars are caused by interference from distant stations being propagated via sporadic-E; these stations are not strong enough to "override" the local stations but do cause the visible QRM. You may also hear some audio distortion as well.

Reception during an E-skip opening can vary in a wild and unpredictable manner. It is not uncommon for the signal from one distant TV station to abruptly and completely fade away and be replaced by the signal from yet another distant station; a few minutes later, another abrupt and total fade will bring the first station viewed back. On FM, stations may "surface" above the jumble for just a few seconds and then be replaced by another station. Since sporadic-E clouds are in motion in the E-layer, the "direction" of an opening can change as the clouds move; it is often possible to track such motion by observing how stations on the same frequency but in different areas fade in and out. However, some E-skip openings are stable; it is sometimes possible (if you are so inclined) to spend a half hour or more listening to an FM station or viewing a TV channel over a thousand miles away.

Sporadic-E is most likely to occur from mid-morning to early afternoon, your local time, and again from late afternoon throughout the early evening hours. Some strong and intense openings begin as early as sunrise and can continue until well after midnight, however. E-skip is also more common the closer one is to the

FM and TV DXing

equator; in equatorial regions, it is a year-round, almost daily phenomenon. Thus, listeners in the southern half of the United States will have an advantage over those in the northern half and Canada.

Sporadic-E is not the only propagation mode for distant TV and FM signals. Another is known as "tropo," which refers to propagation via *ducting* in the Earth's troposphere. The troposphere extends from the Earth's surface to an altitude of approximately six miles. This is the layer of the atmosphere where weather takes place; rain, wind, and storms are all products of the troposphere. Normally, the troposphere has no effect whatsoever on radio signals of any frequency; they simply pass through it. But a very curious effect can be noticed sometimes on signals of above approximately 50 MHz in frequency. Normally, the temperature and percentage of water vapor present in the troposphere decreases with altitude. But sometimes a layer of cool, dry air close to the Earth's surface is overridden by a layer of warmer, more moist air, meaning that at a certain point the air temperature and relative humidity both increase. This situation is known as an *inversion* and is commonly found along weather fronts or along large bodies of water. When inversions are present, signals above 50 MHz can get "trapped" at the point where the inversion is present and prevented from traveling out into space. Those signals then follow the curve of the Earth, often up to distances over 500 miles.

One interesting aspect of tropo propagation is that its effects are more pronounced as the frequency of the signal increases. Thus, the FM broadcast band is far more subject to tropo than are TV channels 2 through 6, and TV channel 13 is more affected than the FM broadcast band. In fact, several TV DXers chase DX signals from hundreds of miles away on the UHF channels thanks to tropo propagation.

Tropo also lasts much longer than sporadic-E propagation. It's not uncommon for an inversion along a slow-moving weather system to last for days. Certain coastal regions experience tropo propagation more frequently than inland regions. For example, tropo is frequently present on the path between the west coast of Florida and the coastal regions of Louisiana and Texas around sunset and in early evening. Similar paths can exist at the same times up and down the Atlantic and Pacific coasts. The signals during tropo openings are also more steady than those from sporadic-E.

Despite its "advantages" over sporadic-E propagation, tropo also presents some difficulties. Its range is usually more limited than E-skip; distances of up to 1400 miles over land have been spanned by tropo, and Hawaii and the West Coast are sometimes linked via this mode. But an average sporadic-E opening will permit greater distances to be covered than tropo. Moreover, tropo openings tend to be more geographically localized than sporadic-E. A distance of only a few miles can make the difference between being "in" an inversion opening or being "out" of it. This applies to transmitting FM and TV stations as well as to DXers. For example, a tropo opening could take place between your area and central Ohio; while you could hear or see stations from that area, you may not be able to receive any stations from eastern or western Ohio. In a E-skip opening, the odds are good that stations throughout Ohio and adjacent states could be received.

A more serious challenge is that successful tropo DXing generally requires more elaborate receiving equipment and antennas than does E-skip. Seldom will the "duct" through which FM and TV signals are propagated during a tropo opening reach the ground; some sort of rotatable outside antenna is usually necessary. The higher such an antenna, the better; by the same token, the best sites for tropo DXing are on hilltops. Tropo signals are often not as strong as those via sporadic-E, so a more sensitive receiver is usually necessary. Many FM DXers use receivers designed for use in audio systems; both FM and TV DXers often use mast-mounted-preamplifiers with their antennas.

Tropo paths generally cannot form over mountainous areas. Thus, listeners in Rocky Mountain states and provinces will find tropo rare, and tropo across the Rockies (such as from California to Colorado or Kansas) is extremely rare. The Appalachian Mountains are not high enough to seriously disrupt a tropo path, so a tropo opening is possible in North America between most points east of the Rockies.

The best time for tropo propagation tends to be in late summer and autumn. A good indication that tropo is possible is hazy or foggy weather; the same inversion that causes such weather can also produce tropo. In addition, shorter-range tropo openings (300 miles or less) can also occur around sunrise as the sun first warms the upper layer of the troposphere before the lower layers; such openings usually vanish within an hour after sunrise.

AMATEUR "HAM" RADIO

Many people get SWLs confused with amateur radio operators or, as they are more commonly known, "hams." Some SWLs have been known to pass themselves off as hams, while others act as if there is a major rivalry between the two groups. Such rivalry is sometimes provoked by the condescending manner some radio amateurs display toward SWLs. These attitudes are fortunately rare; many radio hobbyists (such as your author) are both SWLs and hams and find their enjoyment is double what it would be if they were strictly a ham or SWL!

Amateur radio operators are licensed to *transmit* on various frequencies ranging from 160-meters (1800 to 2000 kHz) all the way up in frequency into the "microwave" region. In the United States, the necessary licenses are issued by the Federal Communications Commission in five different classes, with different operating privileges, upon passing a test on operating rules, radio communications theory, and Morse code for the class of license desired.

So what do hams do on the air? They talk to each other—using CW, FM, SSB, RTTY, SSTV, TV, and even a computer-to-computer technology known as "packet radio"—at distances ranging from across town to the other side of the world. Many hams participate in on-the-air contests, trying to see who can score the most points by contacting as many different stations, countries, states, radio "zones," or even different call sign prefixes in a weekend. Other hams swap QSL cards with other hams they contact, and try to earn awards for contacting all states, over 100 different countries, or even all counties in the United States. (There are over 3000 counties in the United States, and several hams have managed to contact and swap QSL cards with every one.)

There's a more serious side to amateur radio. In many major emergencies (such as the 1985 Mexico City earthquake), ham radio is the only two-way radio link able to handle emergency communications on a large scale. Other hams provide auxiliary communications for special events such as parades and running marathons which can overload a city's normal communications networks. To keep in practice for such situations, some hams participate in networks to deliver free messages. A valuable service performed by hams in tornado-prone areas is "weather spotting." When conditions for violent weather exist, hams travel to scattered observation sites and take along "walkie talkie" ham sets or use the ham stations installed in their cars to report any tornado activity or other dangerous weather (hail, heavy lightning, and the like) to the National Weather Service.

Hams have been involved with satellite communications from the very beginnings of the space age. Many hams tuned in the signals from the first satellite, Sputnik I, in 1957 on its frequency near 20 MHz; radio and TV stations made use of recordings made by hams so that the general public could also hear what the first satellite "sounded" like. Radio amateurs have also constructed and

Figure 9.2 Chuck Andrews, operator of amateur station NI5I, handling emergency communications during the 1985 Mexican earthquake.

orbited (with the assistance of various governments) their own series of satellites for amateur radio communications. The very first satellite designed and built without government money was OSCAR I, which was launched December 12, 1961, as part of the payload of a U.S. Air Force satellite launch from California. "OSCAR" is an acronym for "*o*rbiting *s*atellite *c*arrying *a*mateur *r*adio." The first OSCAR was a beacon transmitter operating near 144 MHz (at the lower edge of the amateur two-meter band). In 1965, OSCARs III and IV were launched. These satellites were active relay stations, listening for signals from hams on the ground and retransmitting them. The first satellite communications between the United States and the Soviet Union were between amateurs K2GUN and UP2ON on December 22, 1965, using the OSCAR IV satellite. Since then, several other OSCAR satellites have been launched, and the Soviet Union has launched a series of "RS" (*r*adio *s*putnik) satellites which operate in the same general frequency ranges and manner as the OSCAR series.

Amateur radio operation involving live operators in orbit has also taken place. Actually, ham radio was represented from the beginning of manned spaceflight, since the first person to travel into space, Yuri Gagarin, was an amateur radio operator. He did not, however, operate from space. That event did not take place until November 1983, when Owen Garriott, W5LFL, operated from the space shuttle *Columbia* using a two-meter "transceiver" (a combination transmitter and receiver in one package) for FM voice communications with amateurs on the ground. Since then, other hams have flown into space and treated earthbound amateurs to the thrill of direct communication with someone in space.

Figure 9.3 During the November 1983 flight of the Space Shuttle *Columbia*, many earthbound hams were lucky enough to contact astronaut Owen Garriott, known to other hams as W5LFL.

Amateur "Ham" Radio

But for many hams the biggest attraction remains the ability to communicate with other people. Hams are drawn from everywhere in the human spectrum, which means they are just as interesting or boring (or sane versus "nuts") as any sample of people drawn from the general population. Tune across the voice segments of the ham bands in Table 9-4, and you'll soon see the truth of this. Some hams seem unable to converse on any topic other than the weather or the equipment they are using (and that suits many like-minded hams just fine), while others have truly interesting experiences to share. Where else can you meet and

TABLE 9-4 Major Amateur Radio Bands

Includes modes which can be used within each band segment.

Band	Frequency	Modes
160-Meters	1800–2000 kHz	CW, RTTY, SSB (usually LSB), and SSTV
80-Meters	3500–3750 kHz	CW and RTTY
	3759–4000 kHz	SSB (usually LSB) and SSTV
40-Meters	7000–7150 kHz	CW and RTTY
	7150–7300 kHz	SSB (usually LSB) and SSTV
30-Meters	10100–10150 kHz	CW and RTTY
20-Meters	14000–14150 kHz	CW and RTTY
	14150–14300 kHz	SSB (usually USB) and SSTV
15-Meters	21000–21200 kHz	CW and RTTY
	21200–21450 kHz	SSB (usually USB) and SSTV
12-Meters	24890–24930 kHz	CW and RTTY
	24930–24990 kHz	SSB (usually USB) and SSTV
10-Meters	28000–28300 kHz	CW and RTTY
	28300–29700 kHz	FM, SSB (usually USB), and SSTV
6-Meters	50–50.1 MHz	CW and RTTY
	50.1–54 MHz	FM, SSB (usually USB), and SSTV
2-Meters	144–144.1 MHz	CW and RTTY
	144.1–148 MHz	FM, SSB (usually USB), and SSTV

talk with a sound technician at the Universal Amphitheater in Los Angeles, a retired exploration geologist who has lived on every inhabited continent, a U.S. Navy submarine officer, a New York cable television producer, and a 15-year old who wants to be a scientist—all in the same evening?

Moreover, amateur radio often melts away racial, cultural, class, religious, and national distinctions. Operating skill and acumen cannot be bought. Status in the form of a higher-class license is open to everyone who can pass the requisite tests. And hams who try to converse with (or "QSO") stations in foreign countries find themselves forced to become somewhat "internationalist" in their outlook and attuned to the political and cultural sensitivities of different nations.

The greatest growth area in amateur radio since the early 1970s has been in a more localized form of communications, however. This is the area of *repeater* operations. A repeater is a relay station which listens for stations on one frequency (the "input" frequency), receives and amplifies any stations that it does hear, and retransmits such stations on a different frequency (the "output" frequency). Repeaters generally operate in the VHF and UHF ham bands, where communications range is normally restricted to the horizon (or "line of sight"). Repeaters are placed in favorable locations, such as the tops of mountains or tall buildings, to increase the working range of VHF and UHF ham stations, particularly low-powered stations in cars or hand-held transceivers. Almost all repeater communications are FM, and it's not unusual with a well-placed repeater for two amateurs, using low-powered hand-held transceivers, to be able to communicate with each other over 100 miles apart.

To earn any of the five classes of amateur licenses currently available in the United States, it is necessary to learn the Morse code and pass a test of your ability to receive typical amateur messages transmitted in Morse. At one time, a knowledge of Morse was a practical necessity for a ham license, since voice communications were still in a highly experimental stage. Today, there is no such need and, in fact, most hams use some form of voice communication (SSB, FM, and so on) rather than CW. Recognizing this, international agreements which once required CW tests for ham licenses now permit issuing licenses without a Morse test for operation at frequencies above 144 MHz; some countries, such as Japan, have gone a step further and now issue licenses without a Morse test for frequencies below 144 MHz. Still, the requirement for a CW receiving test remains, despite several attempts to establish a "no code" license. Several arguments have been advanced for the retention of Morse code as a license requirement, but the basic reasons (usually unspoken) seem to be a "I had to do it, and so should you" attitude among already-licensed hams and a desire among those same hams to limit the number of persons holding amateur licenses. This has resulted in a dramatic slowing in the growth of those holding amateur licenses, particularly among younger people. Many young people with technical interests who in the past might have gone into ham radio as a hobby have been put off by the Morse code requirement and have opted for other interests such as computing. (The designer of the first Apple computers, Steve Wozniak, was a former ham who left the hobby for computing.) Other nations, such as Great Britain, Australia, the Soviet Union, and Japan, have instituted code-free licenses without any adverse effects on amateur radio in those nations. Despite this, knowledge of Morse code remains a requirement for an American ham radio license and likely will for the forseeable future. (Interestingly, the FCC has twice in the past 15 years proposed issuance of no-code licenses for above 144 MHz, but has given up in the face of opposition from already-licensed hams!)

Fortunately, this requirement is not as difficult as it may seem at first. The code speed requirement for the Novice and Technician class licenses is only five words (an average of five characters to each word) per minute. ("Words per minute" is abbreviated "wpm.") This is a very slow speed, meaning that on average one Morse code character (letter or number from 0 to 9) is sent in slightly more than two seconds. Learning Morse code at this speed is about as difficult as learning the letters of the alphabet in alphabetical order and learning to count from one to ten. It's so simple, in fact, that many seven- and eight-year olds have mastered the code at that speed and have received ham licenses. The biggest problem most youngsters at those ages have is not in learning CW, but reading well enough to pass the written tests.

The written tests are all in a multiple-choice format, and are of increasing difficulty by license class. While some study and effort are necessary to pass the written exams, most questions relate directly to amateur radio rules, safety, equipment, antennas, and other knowledge necessary to properly and safely operate a station. Numerous study guides are published which help prepare for the written exams. Most people can pass the Novice class written exam after only a few hours of study, and the remaining classes of written exams should require only a few weeks of part-time study for each higher class.

Once a license is obtained, it is good for ten years and may be renewed free when expiration nears. Actually, a ham radio license consists of two parts known as "operator" and "station" licenses. The operator portion allows the individual named to operate any amateur station up to the extent of his or her license class. For example, someone who holds a Novice license may visit and operate a station belonging to an Extra class amateur, but only under the conditions allowed for a Novice. If the Extra were to visit the Novice, the Extra could operate the Novice's station to the full extent of Extra privileges. The station portion is a call sign indicated for the station at a permanent location indicated on the license; the call sign is also used to identify operation from portable and temporary locations or while "mobile" by car, foot, and so on. In practice, both licenses come on the same form from the FCC and are considered part of the same "ticket" (as ham licenses are known); many hams are better known by their call signs than by their full names.

The first license level is the *Novice* class. It requires passing a 5-wpm code test and a simple 20-question exam. It allows the use of Morse code *only* from 3700 to 3750, 7100 to 7150, 21100 to 21200, and 28100 to 28200 kHz with a maximum transmitter power of 200 W. While restricted to Morse code only, there is no restriction on how this is sent; a hand key may be used, or a hardware/ software interface can be used to send and receive the Morse code with a personal computer.

The next level is the *Technician* class. This license also requires a 5-wpm code test, the Novice written exam, and a 50-question exam on general topics in radio theory and operation. If one already has an unexpired Novice license, full credit for the Novice code and written exams are given, and only the general-level written exam must be taken. The Technician is allowed all Novice privileges plus all amateur privileges, including voice and repeater operation, on frequencies above 50 MHz. Technicians are allowed to use up to 1500 watts of power above 50 MHz. Besides the two-meter band (144 to 148 MHz), another favorite Technician is six-meters (50 to 54 MHz), which has sporadic-E propagation much like TV channel 2.

The "standard" amateur license is the *General* class. Its requirements are identical to the Technician class, except that the Morse code test requirement is 13 wpm. It gives all amateur privileges above 50 MHz and on the 160-, 30-, 12-, and 10-meter bands and privileges on portions of the remaining bands. In practice, this means that Generals have access to about half the frequencies set aside for voice operation on 80-, 40-, 20-, and 15-meters and most CW privileges with the exception of the lower 25 kHz of 80-, 40-, 20-, and 15-meters. Generals and all other higher-class licenses can use up to 1500 W of power except in the ranges

Figure 9.4 Simple interface units allow microcomputers to be used in a ham station for CW and RTTY communications.

Novices are allowed to operate and on 30-meters; everyone is restricted to 200 W in those frequencies.

The *Advanced* class license gives almost all voice privileges; however, some voice segments are still off limits, as are the CW segments located in the lower 25 kHz of the 80- through 15-meter bands. For a holder of a General class license, the only new requirement to qualify is passing a 50-question exam dealing with more advanced radio theory and techniques.

The *Extra* class gives access to all amateur frequencies and operating privileges. For the holder of an Advanced license, it requires passing a 20-wpm Morse code test along with a 40-question test on highly advanced radio theory and communications methods.

The call signs of amateur stations usually reflect the license class held by the operator. All call signs are issued from standard international allocations (see the appendix for the complete list) and consist of a one- or two-letter prefix, a single digit, followed by one to three letters. All Novice call signs begin with two letters, followed by a digit and three letters. For example, a new Novice might have a call similar to WZ6ABC. If this Novice were to upgrade to a Technician or General license, he or she would be eligible for a call sign composed of a single letter followed by a digit and three letters, as in W6ZZZ. Advanced license holders are eligible for call signs beginning with two letters followed by a digit and two letters, as in KM6RR. Extra class licensees may receive a call consisting of a combination of three letters and one digit, as in KZ6X, W6AM, NN6F, and so on. The letters composing a call are assigned in alphabetical sequence, while the digit represents the call sign district in which the amateur was first licensed. The United States is divided into ten such districts numbered from 0 to 9.

Although amateur licenses are issued by the FCC, all exams are administered by amateurs themselves. Currently, any licensed amateur holding a General class or higher license may administer a Novice examination. Exams for Technician class and higher licenses are administered by teams of three or more Extra class amateurs who, under the direction of various regional coordinators, schedule, publicize, and conduct examination sessions. The amateur exam teams prepare and conduct the exams and certify the results; the FCC then issues ham licenses based upon these certifications. Although there is no fee for the license itself, the coordinators conducting exams are authorized to charge an examination fee equal to their expenses; at the time this book was written, the maximum permitted fee was slightly over four dollars.

The national association of radio amateurs in the United States is The American Radio Relay League (ARRL), located at 225 Main Street, Newington, Connecticut, 06111. Approximately one-third of America's 400,000 licensed hams belong to ARRL and receive its monthly magazine *QST*. The ARRL is a nonprofit organization which sponsors a variety of training programs, operating activities, and awards for hams; its numerous publications are valuable additions to any amateur's library. A note to ARRL will bring you membership and publications information.

Tuning across the bands described in Table 9-4 will give you a good idea of the types of activities hams engage in. If your interest is caught, the best step would be to contact a local ham club. Many have training courses to prepare for the various exams; every one will have members willing to help beginners and to demonstrate ham radio. The names and addresses of clubs in your area can be obtained by dropping a note to ARRL and enclosing a self-addressed stamped envelope with your request.

At the time this book was written, the FCC was considering several proposals to add extra privileges for Novice license holders. Among the most prominent of these was to allow Novices to use SSB on 10-meters and to permit using FM and repeaters on some VHF and UHF bands. These proposals may or may not have been adopted; the ARRL will have the latest details.

If you should go on to get your own ham license, be sure to listen for a station identifying itself as KR2H. That's your author, and he'd love to talk things over with you!

10

Unusual, Illegal, and Mysterious Radio Activity

Not everything you can hear on SW falls into the categories discussed in previous chapters. There is a large and growing body of activity which falls outside the usual "bounds" of SW communications, such as clandestine and pirate broadcasting stations, mysterious beacons, transmissions in various languages consisting of nothing more than groups of numbers, illegal transmissions designed to facilitate smuggling or other illegal activities, and some signals which defy categorization. All of these can be heard, often easily, on a general-coverage receiver.

The key to hearing much of the activity in this section is patience. These stations seldom operate on regular schedules and frequencies; instead, you'll have to carefully search for such stations. This means that hearing the types of stations covered in this chapter often depends on "being in the right place at the right time" more than it does upon your equipment or DXing skill. Another important tool is information. Membership in a SWL club or subscription to a commercial SWL publication is essential to keep current on the many rapid changes that can take place in such activity.

CLANDESTINE BROADCASTING

Clandestine broadcasting stations sound much like any normal broadcasting stations. However, deception and illegality are important elements in clandestine operation. For example, a clandestine broadcaster may keep its actual location a secret or pretend to operate from a location different from its true location. The true purpose or sponsor of a clandestine broadcaster is inevitably hidden; moreover, they are always political creatures. Clandestine broadcasters exist to bring about some political change or action in their target countries. When political conditions in the target country or in the sponsoring country change, the clandestine goes off the air or a new clandestine broadcaster may go on the air. However, certain patterns of clandestine operation tend to repeat themselves, with the same old techniques often pressed into service for new causes.

The first clandestine broadcaster took to the air in 1941. This was Radio Espana Independiente, the voice of the exiled Spanish Communist Party. This station first took to the air from transmitters in the Soviet Union and continued to broadcast from locations in Eastern Europe until it left the air in 1977 following the death of Francisco Franco. During its years of operation, it called itself "the station of the Pyrenees" (the mountains on the border between France and Spain) and was well heard in North America thanks to the powerful transmitters it used. A similar station, Radio Portugal Livre, began operations soon after Radio Espana Independiente and continued until the early 1970s.

After World War II, several new clandestines from the Soviet bloc directed toward Italy and Iran came on the air but likewise disappeared in the early 1970s. The United States was also busy with clandestine activity. Radio Free Europe and Radio Liberty both came on the air from West Germany in the early 1950s, broadcasting to the Soviet Union and eastern Europe on SW. Both of these stations claimed to be private operations supported entirely by donations. Many magazines and TV stations in the United States ran public-service solicitations for funds for these stations. But in 1972, it was revealed that both stations were actually operated by the Central Intelligence Agency (CIA). Congressional hearings disclosed that in one year these stations received over $12,000,000 in free advertising yet collected less than $100,000 in donations. Congress voted to continue funding these stations, but through normal congressional appropriations rather than through the CIA.

The CIA was also responsible for the most famous of all clandestine broadcasters, the now-legendary Radio Swan/Americas. In May 1960, SWLs were surprised to hear a station calling itself Radio Swan begin operations on 1160 and 6000 kHz. All programming was in Spanish, with a strong anti-Castro slant, and the station claimed to be a commercial operation located on Swan Island, an island in the Gulf of Mexico between Honduras and Cuba. The potent signals from the station were clear throughout the United States and were soon subject to heavy Cuban jamming.

However, some very peculiar aspects of Radio Swan soon became apparent. Swan Island was claimed by both the United States and Honduras, and the FCC even issued licenses for amateur radio operation there; however, the FCC claimed to know nothing whatsoever about Radio Swan. The station was supposedly owned by a "Gibraltar Steamship Corporation," based in Miami, which turned out upon further checking to own no ships of any description. Further, the few commercials aired seemed to be inadequate to support an operation on the scale of Radio Swan. Fidel Castro himself was convinced the station was an American covert operation; in a September 1960 speech at the United Nations in New York (which lasted over four and a half hours), Castro claimed Radio Swan was operated by the U.S. State Department to undermine his regime.

Any pretense that Radio Swan was merely a commercial station went out the window with the abortive Bay of Pigs invasion in May 1961. Radio Swan transmitted coded messages to the invading forces and guerillas inside Cuba, suspending normal programs during the invasion and serving as a communications link for the invaders. A few months after this, the station changed its name to Radio Americas and continued operating until the spring of 1968. Its programs as Radio Americas consisted of many anti-Castro talks and newscasts, and much pop music in English and Spanish was featured.

There was some controversy concerning this station's actual location and ownership during its heyday, but it is now beyond doubt that the station was on Swan Island and was indeed operated by the CIA. The editor and a writer for the now-defunct *Electronics Illustrated* magazine were allowed to visit Swan Island in early 1968 and observe Radio Americas in full operation. Swan Island was also home to a LW radionavigation beacon with the call sign SWA, and several of the personnel assigned to it and a U.S. Weather Bureau station there publicly stated that Radio Americas was on Swan Island. The station was built by A. D. Ring and Associates of Washington, and Mr. Ring back in 1960 described the facilities on Swan for newspapers such as the Washington *Daily News;* the facilities described were identical to those the *Electronics Illustrated* visitors found in 1968. Even the station's QSL card showed Swan Island as the location.

The links to the CIA were more circumstantial but abundant. For example, the president of Gibraltar Steamship was Thomas D. Cabot, who had once been head of the U.S. State Department's Office of International Security Affairs. The station's manager was Horton Heath, who was the former director of the U.S. State Department's Office of Internal Security. A direct link to the CIA was made by former CIA agent Victor Marchetti in his 1974 book, *The CIA and the Cult of Intelligence*, in which he stated that the CIA had been behind the station from the beginning and that the station was on Swan Island. As a former CIA employee, Marchetti was required to submit the book manuscript to the CIA for review; the agency deleted several sections of the book on the grounds of national security but left the material on Radio Americas intact.

Even though the events are now two decades old, the subject of Radio Swan/Americas still produces shock waves in certain government circles. Your author submitted a Freedom of Information Act (FOIA) request to the CIA for any documents they might have relating to the station, and received a "We cannot confirm or deny that we may or may not have such material" letter in response. An appeal of this ruling was denied by the agency's review board. A similar FOIA request to the FBI did produce documents; however, they were mostly newspaper clippings. Several internal FBI memoranda were heavily deleted, and some documents were withheld altogether. A FOIA request to the FCC produced only a report on broadcasting in Cuba, prepared by a well-known Florida SWL, and nothing else. When your author queried the author of the report on whether he had sent a copy of his report to the FCC, it turned out he had not.

A curious operation during the same period as Radio Swan/Americas was Radio New York Worldwide, which first used the call WRUL and then WNYW. This was a licensed shortwave broadcaster with studios in New York City and transmitters in Massachusetts. Like Radio Swan/Americas, this was supposedly a commercial broadcaster. It transmitted programs in both English and Spanish, and many of the programs (even in English) had a strong anti-Castro emphasis, even to the point where dedications were made to listeners in Cuba and listeners were urged to write to the station using the name of an announcer rather than the station. Some of the station's programs were even relayed by Radio Americas. This cooperative relaying of programs began in September 1960, back in the days of Radio Swan; moreover, WRUL itself began its anti-Castro programs in April 1960, less than a month before Radio Swan first took to the air. Radio New York Worldwide also had its own SWL club and DX program called "DXing Worldwide" in both English and Spanish editions to attract listeners.

Like Radio Swan/Americas, it seemed unlikely that the few ads the station ran could support the station. Many SWLs were baffled in 1966 when the station announced plans to construct larger transmitter facilities and begin operation in several European languages, since it was unclear where the money to pay for these plans was going to come from. However, the plans were cancelled in 1967 when a fire destroyed the station's existing transmitter facilities. Shortly afterwards, the station was sold by Radio New York Worldwide, Inc., to the Mormon Church; they in turn later sold the station to CBS. Both found the station to be a money-losing venture, and in 1972 CBS sold the station to Family Radio, a California-based religious broadcasting organization. Family Radio changed the station's call to WYFR, built a new transmitting facility in Florida, and still operates WYFR at the present time.

Several questions still surround WRUL/WNYW, particularly the degree to which it was covertly funded by the CIA or other agencies. It is obvious that some source of funds other than advertising had to flow into the station, and the close links to Radio Swan/Americas seem unlikely for a supposedly profit-seeking private company. FOIA queries to the CIA concerning WRUL/WNYW produced the same reaction as did a query of Radio Swan/Americas, and FCC files on the station have nothing out of the ordinary in them. A FOIA request to the FBI revealed that the bureau had quite an interest in the station; unfortunately, most of the documents are so heavily riddled with deletions of some sort that they are unintelligible. Figure 10-1 shows a memo to FBI Director J. Edgar Hoover concerning the fire at the WNYW transmitter site; Figure 10-2 shows a typical censored memo concerning the station.

Another "hot spot" for clandestine broadcasting in the 1960s was Asia. The Vietnam War spawned numerous clandestine stations operated by North Vietnam, South Vietnam, and the United States. The best-known was perhaps Liberation Radio, the voice of the National Liberation Front (Viet Cong). Although it claimed to be operating from a hidden location in South Vietnam, it actually used

```
FD-36 (Rev. 5-22-64)

                            FBI

                            Date: 4/13/67

Transmit the following in _____
                              (Type in plaintext or code)

Via      AIRTEL
         _____
                              (Priority)
----------------------------------------------------

TO:      DIRECTOR, FBI

FROM:    SAC, WFO (98-New) (RUC)

UNSUB;
Fire at Transmitter Sites,
Radio New York Worldwide, Inc.,
Scituate, Massachusetts, 4/9/67.
SABOTAGE
(OO:BS)

            On 4/11/67, [redacted]
N. W., Washington, D.C., telephonically contacted WFO and
advised as follows:

             [redacted] the Radio
New York Worldwide, Inc., 485 Madison Avenue, New York, New
York, which corporation is a transmittal shortwave broad-
casting company with five 5000-watt transmitters located at
Scituate, Massachusetts. The corporation is active in broad-
casting pro-U.S. programs throughout the world, primarily to
Europe, Africa and Central and South America.

             [redacted] indicated that broadcasting in the past had
been concerned with anti-Castro programs, which broadcasts
had received some criticism from unnamed sources.

(3)- Bureau
 2 - Boston (RM)
 1 - New York (Info)(RM)
 1 - WFO

   dmb
(7)

AIRTEL
```

Figure 10.1 FBI memo regarding the fire at the Radio New York Worldwide transmitter site.

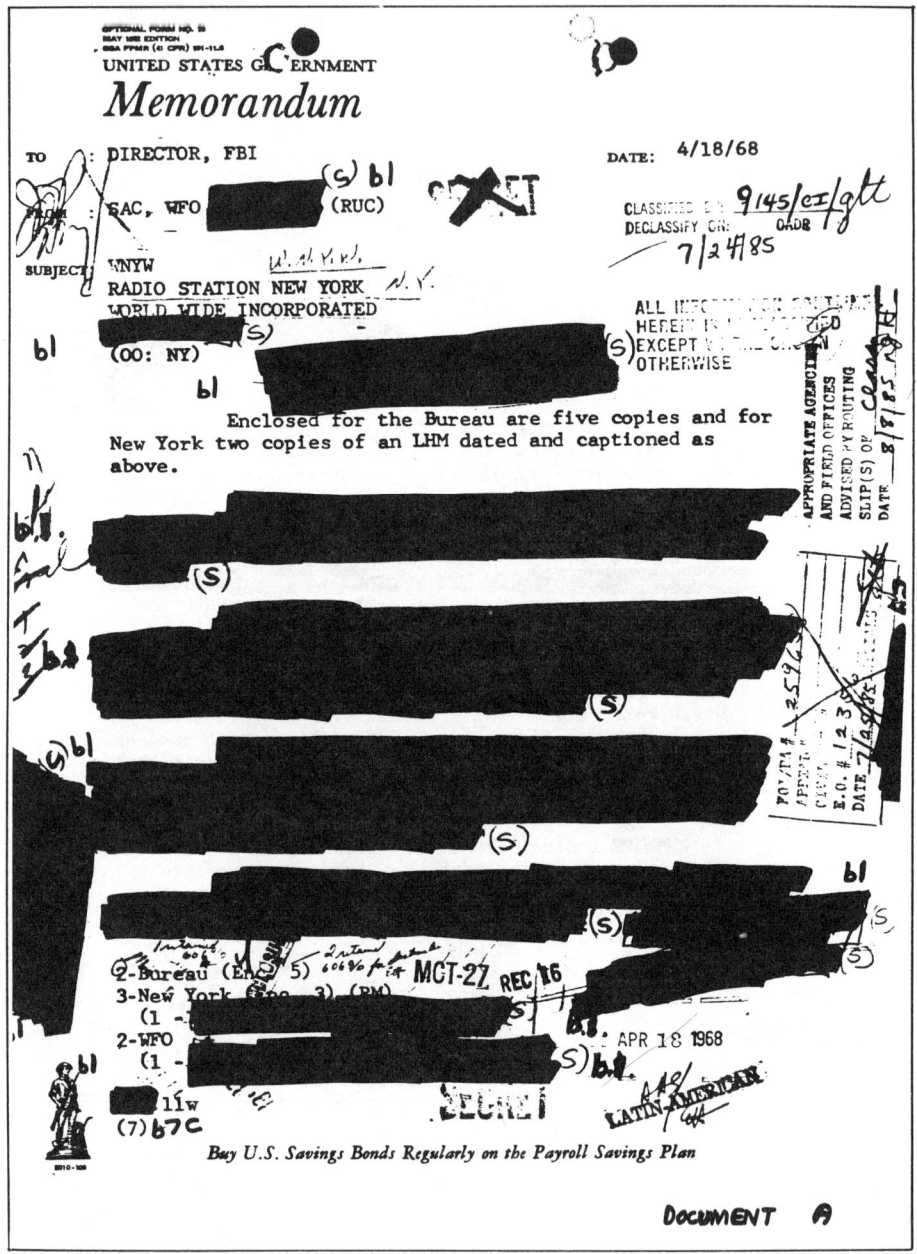

Figure 10.2 An almost completely censored FBI memo regarding WNYW.

the facilities of Hanoi's Voice of Vietnam. Former CIA agent Frank Snepp claimed in his book *Decent Interval* that the agency operated five clandestines, all sharing the same studios in Saigon. All such stations fell silent in 1975 with the defeat of South Vietnam.

Vietnam was not the only arena for clandestines during the late 1960s and early 1970s. In Cambodia, one American operation used devices to alter an announcer's voice so that he could imitate that of exiled Prince Norodom Sihanouk. The 1978 war between Cambodia and Vietnam resulted in Vietnam's establishing a clandestine station opposing the then-ruling Pol Pot government; after the defeat by Vietnam, the exiled Pol Pot regime began broadcasting to Cambodia over a clandestine in China.

Clandestine activity at this time once again revolves around Central America. Like the ghosts of clandestines past, Cuba is once again the target of several anti-Castro broadcasters, while Cuba has turned the tables and now sponsors several clandestines of its own.

Several of the anti-Castro clandestines are sponsored by Cuban exile groups. Among these are La Voz del Cuba Indendiente y Democratica (CID), which is operated by the Cuban exile group of the same name. It has operated several stations which use different slogans. A similar station is La Voz de Alpha 66, run by the famous Cuban exile group of the same name. Additional "private clandestines" have been sporadically active under such names as Radio Mambi, Radio Libertad Cubana, and Radio Antorcha Martiana. Some of these clandestines have operated from southern Florida and have been closed by the FCC; others operate from friendly territory including, apparently, the United States itself.

A major new anti-Castro station began operation in 1985 by playing almost nothing but music. The station became known among SWLs as "Radio Nat King Cole" because of the large number of records played that were in the style of that famous singer. It wasn't until several months later that voice announcements were added, and the station began identifying itself as Radio Caiman (Spanish for "alligator"). The station now has added more contemporary music, special programs for teenage and young adult listeners (such as "La Hora de Juventud"), and slips in anti-Castro messages and quips in a very smooth, "laid-back" style. Unlike the private anti-Castro broadcasters, Radio Caiman has a professional sound, stable frequencies, and no announced backing group. Information has placed the transmitter site for this station in Guatemala; at first glance, it seems as if Radio Caiman borrowed much from the old Radio Americas in its concept and *modus operandi*.

A clandestine which is very similar to Radio Caiman is Radio Monimbo, which opposes the Sandinista government in Nicaragua. Radio Monimbo also uses a "soft sell" approach, with much pop music. The announcing and audio quality of the station is highly professional, as is its frequency stability and adherence to its operating schedule. Reports have placed this station in El Salvador, and it wouldn't be a great surprise to anyone who has listened to both Radio

Caiman and Radio Monimbo if both turned out to be sponsored by the same group or agency.

Another broadcaster opposed to the Sandinistas is Radio Quince (15) de Septiembre, the voice of the Fuerza Democratica Nicaraguense (FDN), better known as the "contras." This station is very "hyper" compared to Radio Monimbo, with many excited shouts, sound effects (including machine gun fire), and strident anti-Sandinista talks. This one has used several different frequencies and transmission times over its history.

In 1979, Cuba entered the clandestine broadcasting wars through a station known as Radio Sandino. This station supported the Sandinista forces, which eventually ousted the Somoza regime in Nicaragua. Although the station tried to give the impression that it was located somewhere in Nicaragua itself, it actually transmitted from a site in Cuba. In addition to frequent pro-Cuban and anti-American commentaries, listeners heard a variety of music (including some of the same selections heard over Radio Havana Cuba), along with directions on how to make a Molotov cocktail and fire an automatic rifle. Radio Sandino continued on the air until the eventual victory of the Sandinistas; today, it is the name used by a station in Managua broadcasting on AM and SW.

The major Cuban-backed clandestine operation at this time is Radio Venceremos, voice of the Farabundo Marti Liberation Front seeking to oust the current government of El Salvador. This station has remarkable similarities to the old Radio Sandino, including some strong rhetoric, revolutionary slogans, and Spanish pop music. In addition to SW, this station claims to broadcast on FM, so a location in Nicaragua is suspected.

South Africa is believed to be home to two clandestines aimed at other African nations. One is A Voz de Verdade, which has been on the air for several years and opposes the Marxist government in Angola. Another is Radio Truth, which broadcasts to Zimbabwe. It might seem that South Africa itself would be the target of clandestine activity; however, other African nations openly make their broadcasting facilities available to groups seeking to overthrow the white minority government in South Africa.

North Africa and the Middle East have much clandestine activity, particularly directed toward Iran. However, these stations are often difficult to hear in North America due to scheduling and frequency choices. One of the better heard clandestines from this part of the world is the Voice of Libyan People, which is believed to be in Egypt.

The two Koreas have also been aiming clandestines at each other for some time. One of the interesting twists here is the attempt by some of these stations to pretend to be based in the capital of the target country. One example is Echo of the Public, a North Korean operation which announces its location as Seoul in South Korea.

Table 10-1 lists some clandestine stations which can be heard in North America without great difficulty. However, the schedules and frequencies of the various stations, as well as their very existence, can abruptly change without any

Pirate Radio

TABLE 10-1 Recent Clandestine Radio Activity

kHz	Station
3670	Radio Venceremos, operated by the Farabundo Marti Liberation Front in opposition to government of El Salvador; sign-ons at 0000, 0200, and 1215 UTC
4950	A Voz de Verdade, directed to Angola around 0300 to 0415 UTC; believed to be located in South Africa
5015	Radio Truth, to Zimbabwe from South Africa at 0430 UTC sign-on in English
5570	Radio Quince de Septiembre, anti-Nicaragua regime station in Spanish, around 0400 UTC
5885	Echo of the Public, Korean-language station pretending to be in South Korea but actually in the North; 1100 to 1400 UTC
6230	Radio Monimbo, anticommunist station to Nicaragua; sign-on at 0000 and 0200 UTC in Spanish
6305	La Voz del Cuba Independiente y Democratica (CID), an anti-Castro station in Spanish, around 0630
6557	Radio Venceremos, same schedule as 3670 kHz
6666	La Voz de Alpha 66, an anti-Castro station in Spanish, sign-on scheduled at 0100 UTC
9940	La Voz del CID, same as 6305 kHz, but at 1430 and 1930 UTC
9960	Radio Caiman, an anti-Castro station in Spanish, with sign-on at 1100 and 0000 UTC
11975	Voice of the Libyan People, believed to be in Egypt, at 0400 to 0600 and 1800 to 2000 UTC in Arabic

warning. As mentioned before, membership in a SWL club or subscription to a commercial SWL publication is the only way to keep up with clandestine activity.

PIRATE RADIO

Like clandestines, pirate stations are secretive broadcasters who often employ a great deal of deception in their operations. Unlike clandestines, however, pirate stations generally have no political ax to grind. They are on the air for the same purpose as your local AM and FM stations—to provide entertainment and amusement. What separates pirates from conventional broadcasters is that pirates operate without a license in total disregard for the radio laws of the country they operate from.

Pirate radio got its start in Europe. In most European nations, broadcasting was (and largely still is) a government monopoly. Radio broadcasting was restricted to a very limited number of government-run networks (such as four in all of Great Britain), which carried the same programs throughout the nation. Moreover, the government-run broadcasters tended to be heavy on classical music, opera, drama, and similar high-brow cultural fare, and advertising was strictly forbidden. The rock and roll music explosion that began in the late 1950s resulted in several nations' establishing so-called "light" or "pop" music programs, with selections running the gamut from Elvis to Sinatra. Such programs sometimes were aired for only one or two hours per day, however. Broadcasting was viewed as a preserve strictly for the government. An example of this attitude was a remark by Britain's Lord Wells-Pestell, who was a member of the Home Office in the early 1960s. Concerning proposals to establish private broadcasting in the United Kingdom, he observed, "I think we have to seriously consider the enormous disadvantages of having a vast army of people who can communicate with one another very easily."

The first European pirates were all commercial ventures operating from ships in international waters such as the North Sea. The first "sea pirate" to go on the air was Radio Mercur, which began operations on July 11, 1958 on 93.12 MHz FM with programs for Denmark. In 1959, another ship station known as Radio Veronica began tests off the coast of the Netherlands; the next year it began regular programs in Dutch, to which the Netherlands government responded by jamming the station.

Such early efforts pale in comparison to the explosion of sea-based pirate stations which began broadcasting to Great Britain in 1964. The first of these British "pop pirates" was Radio Caroline, which signed on at 1200 UTC on March 28, 1964, on 1520 kHz. Announcer Simon Dee opened the station by saying "Hello everybody. This is Radio Caroline broadcasting on 199 meters, your all-day music station" and immediately playing "Can't Buy Me Love" by the Beatles. Radio Caroline was quickly followed by such other sea-based broadcasters as Radio London, Radio Scotland, Radio 390, Radio City, and Britain Radio, all of whom used a format remarkably similar to the "Top Forty" format popular with many American and Canadian broadcasters during that same period. It all came to an end, however, on August 15, 1967, when the British Parliament passed "The Marine Broadcasting Offences Act," which made it illegal to advertise on, supply, work for, or otherwise assist an offshore broadcaster. Most pirates closed the day the act was passed, and the remainder gradually left the air. The major holdout was Radio Caroline, which responded by moving its offices and supply points to Amsterdam. At the same time the sea-based pirates were leaving the air, the BBC announced plans to establish local stations and expand rock and pop music broadcasting. Many of the announcers and disk jockeys who first achieved fame on the various ship pirates eventually found employment with the BBC and local stations. (Some, such as David Lee Travis and John Peel, can even be heard on the BBC World Service today.) True private commercial broadcasting came to Britain in 1971 with the establishment of the Independent Broadcasting Authority, which would grant "franchises" for private broadcasting.

Throughout the late 1960s and early 1970s, a similar pattern was followed throughout much of Europe as more rock and pop music was added to existing stations and limited private and commercial broadcasting was permitted. However, this did not fully satisfy many younger listeners. Part of the problem was that the number of broadcasters remained low, and so the variety of programming was limited as the private broadcasters tried to be all things to all people. (One example is London, which has only two private broadcasting stations for the entire city.) Thus, those who were interested in music such as jazz, soul, and country still found a limited amount of broadcasting for them. Moreover, there were some curious restrictions placed upon the "independent" broadcasters, such as a requirement that no more than 50 percent of the programming could be music. Royalty fees for records played were high, and could amount to several hundreds of dollars per each hour of air time. The old pressures for alternative broadcasting outlets began to build up again, but this time the stations were located on land and operated on a nonprofit basis—almost as a public service—by young people and members of various racial and ethnic minorities.

The new generation of pirates, particularly in Britain, began springing up in the late 1970s to provide programming and serve audiences that existing government and commercial stations did not adequately cover. Unfortunately for listeners in North America, many of these stations operated on FM or on AM with low power, making their reception outside of the United Kingdom difficult. In London, such stations as JFM, Horizon Radio, and Dred Broadcasting broadcast to that city's growing black community (many of them immigrants) with soul and reggae music. London Greek Radio and the Voice of the Immigrants both broadcast entirely in Greek with Greek music for London's Greek immigrant community. Other stations had a more local focus, such as Radio Gemini, which served the sections of London located south of the Thames River. Similar stations also began operations throughout Britain, such as Robin Hood Radio (in Nottingham) and Merseyside Free Radio in Liverpool.

These British pirates went to ingenious lengths to avoid detection and legal problems. Most were unmanned, with remote transmitters linked to studios by low-powered relay transmitters. For example, the transmitter for Radio Jackie was located in an inconspicuous, unattended locked shed atop a hill in southwest London; the actual studios were several miles away.

Other pirate broadcasters came on the air in other countries, and some of these managed to obtain a quasi-legal status. In Ireland, for example, several inconclusive court rulings have allowed private broadcasting stations to operate, even on a commercial basis, as long as the stations refrain from involvement in politics. One Dublin pirate, Radio Nova, managed to convince a trial judge that the purpose of its transmitter was to heat the room in which it was located; since no one observed the station actually transmitting, the judge ruled Radio Nova innocent of broadcasting without a license. Similar confusion over the proper interpretation of radio laws has resulted in numerous private stations in Italy. Other nations, such as the Netherlands, have adopted a live and let live attitude toward pirates. Some of these stations have used SW frequencies and have been heard, and verified, by SWLs in North America.

By 1984, British authorities started a campaign to shut down land-based pirates. They were highly successful in this effort, closing Radio Jackie and numerous other stations. This campaign seemed to be in response to renewed sea-based pirate activity from Radio Caroline (still carrying on after two decades) and Laser 558. The latter station was the source of considerable controversy, as it was an American-based operation, staffed with American announcers, but its sources

Figure 10.3 QSL card from Irish pirate station Radio Dublin on 6910 kHz.

of funding were obscure. The format was rapid-fire music, as epitomized by its slogan, "Never a minute from music on Laser 558." The station initially carried no advertisements, and rumors about its financing ranged from wealthy Irish businessmen to the CIA. The station had a major impact on independent stations in southern England; one announcer at London's Capital Radio stated that the station had lost over 30 percent of its audience to Laser 558.

But Europe has not been alone in pirate broadcasting. The United States and Canada have seen increasing pirate activity in recent years. Unlike Europe, however, North American pirates (most of which are American) seem to have more of a hobby flavor, with many of the participants involved mainly as a lark or for their own entertainment. To be sure, there are some serious undercurrents, primarily through the efforts of some pirate operators and supporters to establish a "citizen's broadcasting band" for low-powered, noncommercial private broadcasting. For most pirates, however, the emphasis seems to be on having fun. Indeed, the impression one often gets is that you're eavesdropping on someone's party. While pirate operations are clearly illegal and cannot be condoned, it's also true that some of the most interesting, outrageous, and genuinely creative things to be heard on radio are broadcast over North American pirates.

In Europe, land-based pirates were inspired by pirates based on ships. That wasn't the case in North America. In fact, there has been only one semi-serious attempt at offshore broadcasting in recent history. The conservative minister Carl McIntire, known for his "Twentieth Century Reformation Hour" program, equipped a ship with an AM transmitter operating on 1160 kHz. On September 19, 1973, this station, calling itself Radio Free America, began operations in international waters off the coast of New Jersey. It was widely heard throughout the northeastern United States for ten hours until the wooden deck of the ship began to smolder as a result of the transmitter's operation. The station left the air and the ship returned to port. That marked the entire history of Radio Free America; the Rev. McIntire later sold the ship and equipment and abandoned his plans for offshore broadcasting.

There was very little land-based pirate activity in North America prior to the early 1970s. Most of the stations which did operate prior to that period did so erratically or were quickly closed down by the FCC or Canada's Department of Communications. The precise factors which resulted in the explosive growth of pirate radio beginning in the late 1970s are difficult to determine, but two major ones seem to have been the availability of old AM amateur radio transmitters at bargain prices and a perception that the FCC was unable to enforce its regulations, as demonstrated by the chaos that swept over the CB channels in the late 1970s. Potential pirates began to perceive, not entirely without good reason, that they could operate with relative impunity as long as they avoided interference to other stations.

Pirates since the 1970s have tended to congregate in two unofficial "pirate broadcasting bands." One is the range just above the standard broadcasting band from 1600 to 1650 kHz; this range is particularly favored by pirate broadcasters in the New York City area. A second band extends from 7300 to 7500 kHz and is used by pirates throughout the United States and Canada. Most transmissions are in AM, and the usual operating times for pirates is after 0400 UTC on weekends.

One of the most fascinating aspects of current North American pirate activity is the many links that these stations have to the SWL hobby; many are operated by SWLs. (Many SWLs who have reported reception of pirates in various SWL club bulletins, but not directly to the station, have been surprised to find QSLs arriving from those stations a few days after the club bulletins were mailed.) The prototype for such stations was the "Voice of the Voyager," which exploded on the scene in early 1978. This station was located near Minneapolis, Minnesota and was operated by several teenage SWLs. It chose 5850 kHz as its frequency and began regular late Saturday night broadcasts which quickly attracted a cult following among SWLs in North America. In addition to the inevitable rock music, the operators took phone calls from SWLs, ran satirical items about the FCC and SWLing, and even gave out the latest SWLing and DX news. This 100-watt pipsqueak likely had a larger, more loyal audience in North America during 1978 than most major international broadcasters!

Unfortunately for its operators, the Voice of the Voyager was raided by the St. Paul, Minnesota office of the FCC in August 1978, and the station was taken off the air. The FCC only issued a warning to the operators, however, and the tone of the "bust" was surprisingly cordial; the FCC agents requested—and received—Voice of the Voyager QSL cards!

The station did not remain silent for long, however. It returned to the air and continued to operate sporadically well into the early 1980s, leaving the air for lengthy periods due to equipment problems and a shortage of participants. The FCC raided the station a second time, and on this occasion there were no warnings or requests for QSLs—instead, fines of several hundreds of dollars were handed out to each of the major participants.

While the Voice of the Voyager itself was eventually silenced, the seed had been planted, and pirate broadcasting was off and running in North America. The FCC and Canada's Communications Canada have made attempts to stem the tide by periodic crackdowns involving hefty fines and confiscation of station equipment. All of these efforts have been heavily publicized. In spite of this, pirates still continue to operate and will likely continue to do so in the future, since the FCC cannot monitor every single frequency and transmitting equipment can be freely bought and possessed even by those lacking an appropriate station or operator license.

The entire issue of pirate broadcasting has become controversial within the SWL hobby and in SWL clubs, with listeners divided on whether acknowledging that these stations even exist encourages them. A few SWL clubs have responded by banning any information about pirates beyond the basic details (time, frequency, and so on) concerning when and where they can be heard and by ruling that QSLs from such stations do not count toward club awards and certificates. (This policy is not followed in regard to European pirates, which are every bit as illegal in the countries from which they operate.) Other clubs are more tolerant, and one specialty club detailing such activity has been established.

Whatever your feelings, pirate broadcasting seems to be here to stay in North America, and many of them are challenging DX. Indeed, trying to hear and extract a QSL from an illegal, low-powered station can be a major challenge. Membership in a SWL club which covers pirate activity is a must; in addition, there is no substitute for careful tuning and listening for pirates, since many are one-shot (or perhaps two-shot) affairs. Table 10-2 lists some recent (at the time this book was written) activity.

"NUMBERS" STATIONS

For years, SWLs have stumbled across transmissions that consisted of little more than a voice, usually that of a woman, reading four- to five-digit groups of numbers in such languages as Spanish, English, and German. These have been the subject of much speculation in the hobby, but today the evidence is overwhelming that such transmissions are actually coded messages to various espionage agents operating in different countries. These stations have become known as *numbers* stations by SWLs.

It doesn't take a great deal of effort on your part to hear these stations. If you tune outside the broadcasting and amateur bands from approximately 0000 to

"Numbers" Stations

TABLE 10-2 Some Typical American Pirate Broadcasters

kHz	Station
1620	"WNYT," rock music and talk around 0500 UTC, believed to be in New York City area
7425	"Voice of Laryngitis," rock music and comedy sketches around 0200 UTC
7430	"Zeppelin Radio Worldwide," German music and announcers with fake German accents around 0400 UTC; believed to be in Ohio
7435	"Secret Mountain Laboratory," country and bluegrass music around 0130 UTC
9470	"Radio Clandestine," rock music and comedy sketches around 2200 UTC; this name has been used by many other pirates in the past.

0800 UTC in North America, you'll run across several such stations, primarily in Spanish. While you can hear numerous stations using four and five digits, the five-digit stations are somewhat more plentiful. Other stations can be heard in English, German, and "tone keyed" Morse code. The latter stations really stand out, since they are AM stations over which Morse code is sent using audio tones. Such stations are heard without using your receiver's BFO. There's a slight variation of the "normal" five-digit pattern, consisting of stations transmitting five-digit blocks but with a slight pause between the third and fourth digit of each group; this pause lasts for approximately the time it would take to give one digit. It's not clear at this time whether this "split group" effect is deliberate or an insignificant quirk.

Numbers stations almost always use a female voice to read the numbers, although on rare occasions a man's voice is used. (It's not known if there is any special significance attached to the use of a man's voice.) Most transmissions are in AM, although some SSB is used, particularly on German-language stations. You'll also notice that the various numbers within each group have the same inflection and sound; a "seven" in the first group of the message will sound exactly the same as a "seven" in the last group. This is because these messages are composed from a recorded set of the digits from 0 to 9 along with a few words used to open and close each transmission. The effect is very similar to listening to a telephone recording for incorrect or discontinued numbers, and very likely the same or similar equipment is used to prepare numbers stations messages. In rare cases, a few apparently "live" numbers transmissions have been heard (once, an announcer paused to cough), but these are very much the exception.

Where do these stations transmit from? It's now apparent, from a variety of sources, that the stations which transmit numbers in blocks of five digits are located in a Soviet bloc country, such as Cuba, East Germany, or the Soviet Union itself. Transmissions of numbers in groups of four digits appear to originate mainly from a site outside of Washington, D.C. However, it is quite likely that other sites have been used by both the so-called "five-digit" and "four-digit" numbers stations.

Several SWLs have attempted to decode these various messages, but this is an excercise in futility. The encoding system used is the *one time pad* method. In this system, the intended recipient of the message has a copy of a pad with columns of numbers, in groups of four or five digits, printed upon it. At the beginning of the message, the sender transmits the *key* to using the one-time pad. The key indicates which page, column, and line the recipient should turn to. The key also tells the recipient whether the number groups in the message should be added to or subtracted from the number groups printed on the pad. After the message is transmitted, the recipient takes the copy of the message and performs the required addition or subtraction to the number groups contained in the one-time pad to get the actual message. Each number group usually stands for a segment of the message, such as "at noon," rather than each one or two digits standing for a letter of the alphabet. To confuse unintended listeners, meaningless number groups can be inserted into the middle of each message. This system sounds simple, and it is compared to some systems. However, its security is close to 100 percent unless copies of the appropriate one-time pad are obtained by outsiders. The principal problem with the method is the time it can take to decode the message; encoding is simple thanks to computer-based systems.

The two "brands" of numbers transmissions (four-digit and five-digit) sign on in different ways for Spanish messages. For five-digit groups, the opening is usually the word "atención" followed by a three-digit group and a two-digit group, as in "atención 545 49." The meaning of the three-digit group is unclear; it could be the intended recipient of the message or the key used to decode the message. The two-digit group is always the number of groups composing the message, so in this example there would be 49 five-digit groups making up the messsge. This sign-on "announcement" is repeated for several minutes before the actual message begins. After the message is transmitted, the words "fínal, fínal" end the message.

Four-digit Spanish stations normally sign on with a three-digit group followed by a count (in Spanish) from 1 to 0; this pattern is repeated several times. The purpose of the three-digit group is uncertain, but it is likely the intended recipient or the key to the message. After this opening pattern has been repeated, the next item is the word "grupo" followed by the number of groups making up the message. This is usually repeated. The four-digit blocks making up the message follow, and the entire message is normally sent twice. Four-digit stations leave the air as soon as the message is sent a second time without any sort of announcement.

German numbers stations follow similar formats, with five-digit stations using "achtung" in place of "atención," while four-digit transmissions open with a three-digit number followed by a two-digit group count. German stations also have a habit of using sound effects (such as tones or beeps) or music at the beginning of transmissions, much like interval signals. Similar patterns also exist for English and tone-keyed Morse numbers stations.

At first, it might seem that numbers stations would operate in an erratic and

"Numbers" Stations

unpredictable manner. Such is not the case, however. Certain frequencies, such as a few kHz above or below 5810 kHz, have been used for decades by numbers stations. Transmissions are often repeated within an hour of the first transmission. (One startling discovery is that some messages are completely repeated, digit for digit, days and weeks after they are first transmitted.) Moreover, many transmissions, particularly of the five-digit variety, tend to be found in certain frequency ranges, depending upon the three-digit key or identifier and two-digit group count at the opening of the transmission.

Table 10-3 is a sampler of numbers stations activity. While times and frequencies change, the items in Table 10-3 are typical of the patterns of such stations. As mentioned before, the best way to hear these stations is to tune carefully outside the broadcasting and amateur bands. The signals are often quite loud, particularly from Spanish stations. In fact, if you're an active utility DXer,

TABLE 10-3 "Numbers" Transmissions

kHz	Description and time (all in UTC)
4025	Five-digit Spanish groups at 0300
4575	Five-digit CW groups sent by tone keying at 0400
4730	Four-digit Spanish groups at 0215
5734	Five-digit German groups at 0345
5090	Split five-digit English groups at 0000
5210	Five-digit English groups at 0115
5316	Split five-digit German groups at 0000
5748	Five-digit German groups at 2315 in USB
5810	Four-digit Spanish groups at 0530 (active frequency)
5975	Five-digit CW groups sent by tone keying at 0645
6405	Five-digit Spanish groups 0500; male announcer
6573	Five-digit Spanish groups at 0815
6770	Five-digit Spanish groups at 0700
6810	Four-digit Spanish groups at 0500 (active frequency)
6840	Five-digit English groups at 2220; sometimes split groups used
6910	Five-digit Spanish groups at 0440; male announcer
7520	Five-digit CW groups sent by tone keying at 0430
7590	Split five-digit English groups at 2015
7846	Five-digit Spanish groups at 0730 in USB
7885	Five-digit Spanish groups at 0730
7910	Five-digit Spanish groups at 0810
8420	Four-digit Spanish groups at 0400
9120	Split five-digit German groups at 2100; sometimes in USB
9222	Split five-digit English groups 0510
11360	Four-digit English groups at 1500
11535	Four-digit Spanish groups at 0215 (active frequency)
11545	Five-digit German groups at 0115 in USB
12240	Five-digit Spanish groups at 0600
13387	Four-digit English groups at 1415 in USB
14588	Five-digit Spanish groups at 2100
15075	Four-digit Spanish groups at 0230

you'll run across many numbers stations without having to try to hear them.

Evidence for the locations of these stations has been assembled from a variety of sources. Some of it is highly reliable, while in other cases it is much less so; some of the evidence is also contradictory. However, it is possible to make some reliable statements about where most are located. It now is clear that all of the five-digit German-language transmissions and most of the five-digit English messages originate from within East Germany, with the facilities at Nauen (used for standard time and frequency station Y3S on 4525 kHz) the most likely site. Evidence for this comes from monitoring from Europe and a virtual "on-site" inspection by one enterprising SWL; in addition, numerous spy trials in Europe have revealed how persons in the West working for intelligence agencies in the Soviet Bloc have received instructions via radio using the one-time pad system. In each case, the one-time pads found have been based on a five-digit system.

The source for the Spanish five-digit transmissions has now been clearly established as Cuba. In the past, the FCC has sometimes replied to queries about the five-digit Spanish stations by stating that they are in Cuba. However, in other cases the FCC has disclaimed any knowledge whatsoever about them. (Your author queried the FCC about five-digit Spanish transmissions he was hearing on 3060 and 3090 kHz in 1978 and received a reply from the FCC's Atlanta office that those frequencies were assigned to the aeronautical mobile radio service, and the signals could be coming from any South or Central American country!) In 1984, your author wrote a monthly column for *Popular Communications* magazine and suggested that readers try calling the FCC's monitoring stations whenever they heard something unusual and try to get the personnel there to identify and locate the source of such signals. A list of FCC monitoring stations was given along with their telephone numbers. Several readers tried calling the FCC while a numbers transmission was taking place, and got nowhere. (One reader who called was advised that he should read my column in *Popular Communications* for information about such stations.) However, one reader happened to know some personnel who worked at an FCC monitoring station, and they performed some "unofficial" direction-finding work for him on several numbers transmissions. They found that the Spanish five-digit stations indeed were located in Cuba, and that the Spanish four-digit stations were coming from a site near Washington, D.C.!

The final location for the four-digit Spanish stations was determined by a mathematics professor at a New England college who visited one likely site armed with a portable receiver. This site was south of Washington, D.C., near Warrenton, Virginia, on the Vint Hill Farm Military Reservation. He discovered that four-digit transmissions were indeed originating from there, along with transmissions of station KKN50 operated by the U.S. State Department.

However, there may be other sites which have been used for Spanish four-digit numbers transmissions. In 1978, your author wrote an article on numbers stations for a national electronics magazine and received a letter from a reader in southern Florida. This reader claimed that he, in a manner similar to the New England professor mentioned above, had used a portable SW receiver to track down the location of some four-digit numbers transmissions, and that he had located a site in southern Florida. Unfortunately, this reader did not come forward with more details or specify the exact location he claims to have located.

It should be clear by now why following numbers stations has such an attraction for many SWLs (including your author). How often does the average person have a chance to match wits with the CIA and KGB, or live out a James Bond-type fantasy? SWLs in such areas as southern Florida or the greater Washington, D.C. area who have portable SW receivers and some patience could likely resolve some of the remaining mysteries surrounding these stations.

MISCELLANEOUS UNUSUAL SIGNALS

There are many other signals heard on SW which are of mysterious origin and purpose. Table 10-4 lists a sampling of some which were heard in the months during which this book was written; it is certain that other mysteries will be on the air by the time you are reading this.

Not every strange signal you hear is connected with espionage, smuggling, guerrilla wars, or similar sinister purposes. Many are the product of the U.S. military services, since such stations can (and do) operate wherever they please

TABLE 10-4 Miscellaneous Unusual Transmissions

kHz	Description and time in UTC
3415	"Alpha Romeo Tango Two" repeated at 0500
4607	"78KLP" and "78HVY" in CW around 0200
5307	"D" CW beacon at 2300
5308	"Z" CW beacon at 0045
5440	"SBM DE 1FR" in CW at 0315
5565	"J3R" sending a "VVV" series in CW at 0200
6210	Smuggling frequency in USB
6595	Several stations exchanging number and letter groups in SSB at 0000 to 0300
6840	"Echo Kilo India Two" repeated at 0200
7446	"Kilo Papa One" repeated at 0200; "Kilo Papa Two" repeated at 0315
7906	"K" CW beacon at 0100
8648	Assorted sound effects (footsteps, beeps, explosions, and the like) in SSB around 0345
8780	Flute-like tones at 2030 in USB
11591	"QRA DE M4Z" in CW around 0115
11593	"G6C" in CW with coded groups around 0115
12747	"Mike India Whiskey Two" repeated at 1915
13450	"62 DE 27" and "26 DE OA" in CW around 1300
14534.6	Various audio tones interspersed with data bursts at 1730
17016	"C" beacon in CW at 1945

without FCC control. Sometimes these stations will pop up in the middle of an international broadcasting band to conduct training exercises! Many of these stations will try to simulate combat conditions, and the results can be easily misinterpreted. For example, in 1984 several SWLs heard rock music being transmitted on the same frequencies as Strategic Air Command traffic in an apparent attempt to jam its operations; it turned out this was part of a plan to train SAC personnel in handling the sort of deliberate QRM they would likely encounter in actual combat situations. And many utility SWLs can tell when reservists are taking part in their annual active duty, since strange signals can be heard due to misuse of and unfamiliarity with military radio equipment.

Such caveats aside, it is also true that there is an enormous and growing amount of unusual, unknown, and shadowy activity taking place throughout the SW spectrum. Part of this is due to the technology of SW reception and transmission. High-performance SW receivers, smaller than a typical hardcover book, with direct-frequency readout and SSB capabilities are now available for about $200.00. The advances in transmitting technology have been perhaps even more dramatic. Many amateur radio transceivers today are capable of operating *anywhere* from 150 kHz to 30 MHz with extremely simple modifications. (In the case of one very popular unit, this involves merely moving an internal switch to the opposite position; in other instances, the necessary modifications mean only clipping or unplugging a couple of wires.) There is no requirement for the purchaser of amateur transmitting equipment to actually have an amateur license. Thus, anyone with $1000.00 or less and the ability to use a screwdriver and wire clippers can easily have a transceiver capable of AM, SSB, or CW operation anywhere in the SW spectrum. This has resulted in a dramatic increase in the use of SW radio by drug smugglers, guerrilla groups, pirate broadcasters, and similar people not likely to go through the formalities of a license, call signs, or established operating procedures.

Smugglers are some of the prime users of this new transceiving technology. Smuggling transmissions are likely to be heard anywhere throughout the SW spectrum, but the 6200 to 7000 and 7300 to 7500 kHz ranges have been particularly active. Most smuggling transmissions heard in North America take place during the hours of darkness, and involve narcotics brought in along the Atlantic and Gulf of Mexico coasts. You'll seldom hear an explicit reference to drugs (although your author once heard a station talking about the load of "brownies" it had); instead, you'll hear remarks about meeting points, arrival times, dropping off items, and similar topics. Smuggling transmissions can be identified by their lack of proper radio procedure (such as lack of call signs and use of whistling), cryptic nature, and operation in incorrect frequency segments. You'll hear both English and Spanish used.

Not all modified transceivers are used by smugglers, however. Many persons are using them in a quasi-amateur radio fashion to stay in touch with friends across the country and world. The 6700- to 7000-kHz range seems to have many of these stations. Some of these operators have formed groups which issue call

Miscellaneous Unusual Signals

signs to identify the various stations, and they can sound much like an ordinary amateur radio contact when you first tune in. Other groups which use modified amateur radio gear are the various guerrilla organizations in Central and South America. Unfortunately, you'll have trouble understanding these communications unless you are fluent in Spanish, but you'll have little trouble running across them. Your author remembers seeing a photograph of Eden Pastora, the leader of the Nicaraguan contras, in a national magazine. Pastora was seated in front of an amateur transceiver with which, the accompanying story went, he communicated with his troops and argued with radio officers in the regular Nicaraguan army!

The increasing number of SWLs who can copy Morse code, either "by ear" or with an interface device for a personal computer, has resulted in a number of unusual CW signals being discovered. Most of these signals are communications between two stations, but unusual call signs (such as 62, 78KLP, 1FR, and the like) are used, and the communications seem to be encoded (involving random blocks of letters and digits). Table 10-4 lists several of these. Some are obviously military of some sort, with the bogus call signs apparently the CW equivalent of a SSB tactical call sign. The purpose of some of the other stations is not clear. These latter stations often employ poor radio procedure and the Morse is sent very sloppily, as if a hand key is being used instead of an electronic keyer. Your author has observed some of these mystery CW stations beginning operation after the end of a Spanish five-digit numbers transmission on nearby frequencies.

A persistent mystery has been numerous single-letter beacons on CW. These stations do nothing but transmit a single letter, such as D or Z, continuously. Often these beacons are seemingly "paired" with another beacon, transmitting a different letter, located a few hundreds of Hz above or below it. In addition, the same letter may be heard on several different frequencies at once. Some listeners have reported that the spacing between letters transmitted and rate at which they are transmitted vary. Most sources indicate that these beacons are operated by the Soviet Navy for some purpose. Propagational factors and rough direction-finding have placed most of these beacons somewhere in the Asiatic portions of the USSR. At least one such beacon has been traced to Cuba. This beacon transmitted the letter "W" on 3584 kHz, in the 80-meter amateur band, and was the subject of several complaints by American amateurs to the FCC. The FCC was able to locate the beacon in Cuba by using its direction-finding facilities, and the beacon was removed from 3584 kHz after a formal American protest. The precise purpose of these beacons, as well as all sites they transmit from, still remain unknown.

A final class of mysteries involves stations which transmit groups of letters in AM from the international phonetic alphabet (see the appendix) using a woman's voice. These stations repeat an identifying phrase, such as "kilo papa two," continuously for hours at a time; these are sometimes interrupted by messages made up of groups of letters from the phonetic alphabet. An article in *Popular Communications* magazine claimed that these stations are used by Mossad, the Israeli intelligence service, to send instructions to its agents.

The key to hearing stations such as these is again patience. Careful tuning outside the "normal" broadcasting and amateur bands will reveal all sorts of surprises. Membership in a SWL club that covers such activity is also helpful in finding out which frequency ranges are "hot" and in determining patterns of activity based upon the receptions of many SWLs.

"OFFICIAL" INTEREST IN THE SHORTWAVE HOBBY

One question many new SWLs have is whether or not government agencies keep track of SWLs who write letters to stations in unfriendly nations and whether writing such letters—or even listening to such stations—can cause problems for SWLs. The answer to both appears to be no. The U.S. government simply lacks the personnel to keep track of all letters written to a station such as Radio Moscow (although it has noted letters from a few SWLs). Moreover, the various agencies involved with issuing security clearances appear to understand that most SWLs are more interested in QSLs than ideologies. For example, your author has received security clearances (up to the secret level) required for various technical writing assignments despite having written extensively on the topics in this chapter.

However, this does not mean that various agencies and organizations are not interested in what SWLs may be hearing.

Various agencies do read and note items in SWL publications. For example, a columnist on pirate radio stations for a SWL publication once received a visit from FBI personnel. The editor of a commercial SWL publication has likewise received several visits from FBI representatives. As noted earlier in this chapter, the FCC also takes note of some SWL publications. Your author was told in a telephone conversation with an FCC employee that some of his previous writings on the topics in this chapter have been read by several FCC personnel.

In an attempt to determine the interest of various agencies in SWL activities, your author submitted numerous Freedom of Information Act requests to those agencies asking if they had files on various SWL clubs and publications. The CIA refused to confirm or deny whether it had files on SWL clubs or publications, and rejected an appeal of that ruling.

The FBI responded to a similar request in a different manner, however. Figure 10-4 shows the FBI's interest in an article, published in the now-defunct *Electronics Illustrated* magazine, telling how to listen to stations operated by agencies such as the FBI and CIA. Interestingly, the *Electronics Illustrated* article concerned itself only with information on file with the International Telecommunications Union, and thus on the public record. Figure 10-5 shows a heavily censored FBI memo regarding the American Shortwave Listeners Club, a California-based SWL club that has been in operation since 1959. The subject of interest to the FBI is impossible to determine from this memo, but no criminal

"Official" Interest in the Shortwave Hobby 211

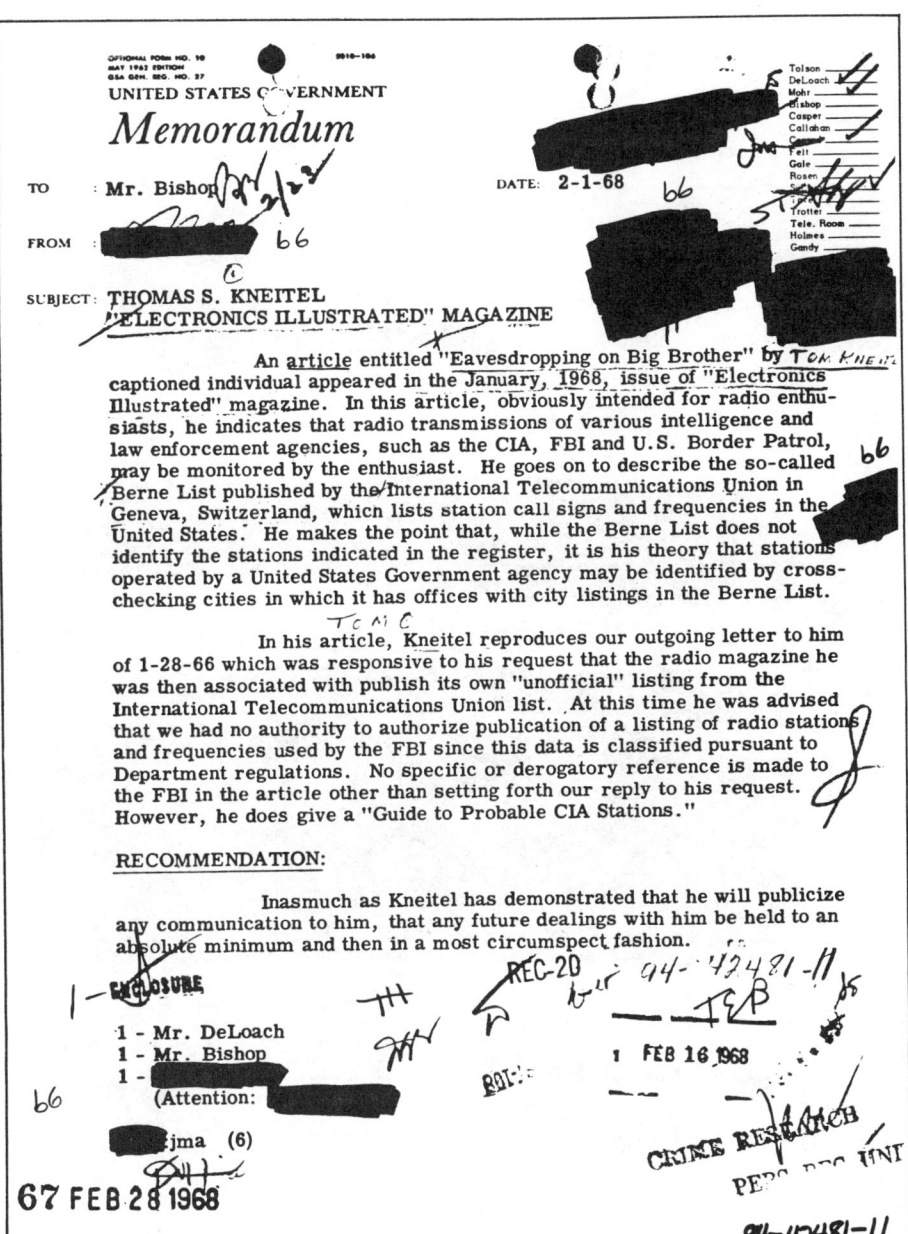

Figure 10.4 FBI memo reacting to a SWLing article in *Electronics Illustrated*.

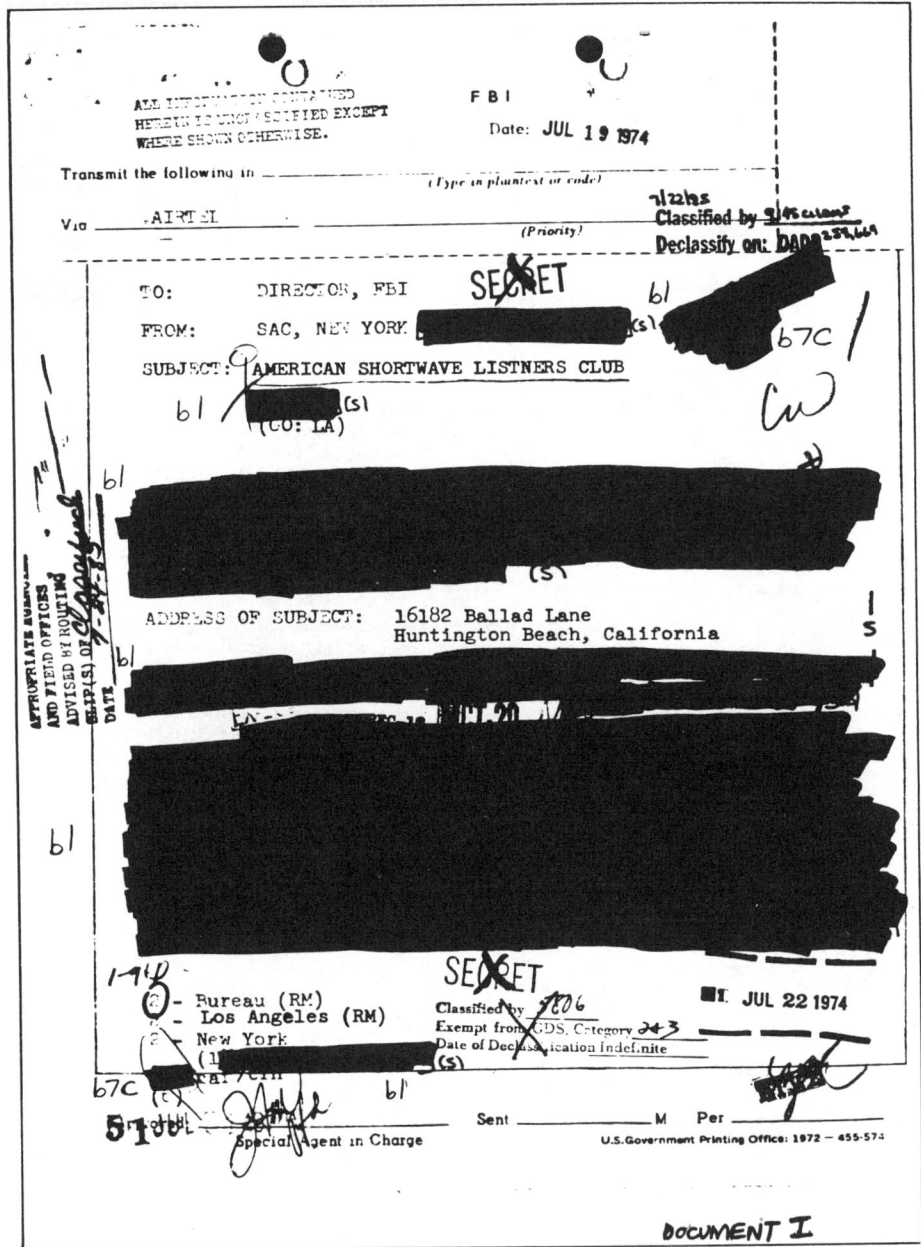

Figure 10.5 Heavily-censored memo dealing with the American Shortwave Listeners Club. The matter it discussed is unknown and cannot be determined.

charges were ever filed against the club or any of its officers. So why was it the subject of a memo classified "secret?"

Whatever the reason, as SWLs develop the capabilities to hear more and more unusual and mysterious signals, it is likely that official interest in SWLing will increase.

11

THE HOBBY OF SHORTWAVE LISTENING

Throughout this book, frequent reference has been made to the hobby aspects of SWLing. The hobby came about initially because of the specialized and esoteric nature of SW reception; the barriers to using and understanding SW radio were so immense that all but the most committed individuals were soon discouraged. Today that's no longer the case, since contemporary SW radios are no more difficult to use than other consumer electronics items. As such, there's really no need to join a SWL club or subscribe to a commercial SW publication if your major interests are restricted to listening to major international broadcasters. Such stations will gladly put you on their mailing lists for program schedules, and will announce changes in their schedules and frequencies over the air several weeks in advance. The only SWL-related publication you'll need in such cases is a current copy of *The World Radio TV Handbook*, which contains schedules, frequencies, and addresses for major SW stations of the world, along with much other useful SWL data.

But many who tune SW radio eventually find themselves joining one or more SWL clubs and/or subscribing to a commercial SW publication. One reason is that information on certain subjects, such as pirate or clandestine radio, is not contained in *The World Radio TV Handbook* and changes rapidly; SWL club bulletins or commercial publications are the only way to keep up to date with what's going on in those areas. Moreover, SWL clubs and publications offer a wealth of information about SWLing and DXing. In their pages you can find information on new equipment, antennas, propagation, new stations on the air, QSLs received, profiles of stations, and listening hints from experienced SWLs and DXers. Finally, many SWLs and DXers enjoy the opportunity to be in contact

Reporting Reception and QSLs

with others who share their interest in SW reception, and to swap information and listening experiences with others.

Perhaps the most common SWL hobby activity is collecting QSL cards and letters from various stations in response to reception reports. Learning how to report reception of various stations can be thought of as the basic skill of the SWL hobby.

REPORTING RECEPTION AND QSLS

Entire books could be written on the subject of reporting reception to collect QSLs (and, in fact, one excellent book has been written on the subject). The techniques covered in this section will be adequate for most international broadcasters as well as AM, FM, and TV stations.

The entire process of reporting reception for a QSL is a transaction between the listener and the station. Both has something the other wants; the listener has observations and comments about what he or she heard, while the station has a card or letter to send the SWL. If the SWL gives the station what it wants, the station can give the SWL what he or she wants. The purpose of a reception report is to enable this exchange to take place. The SWL must prove that the reception, in fact, took place and offer some information to enable the station to do its job better.

Not all SWLs care to collect QSLs, and by the same token not every station cares to send them out. As noted earlier, many international broadcasters have cut back on sending out QSLs, and several persons connected with various international broadcasters, and even some SWLs, have been critical of the entire practice as a waste of limited station resources. (However, most international broadcasters have discovered what American and Canadian broadcasters have known for years—some sort of "promotion," whether a contest or sending QSLs, is usually necessary to build audience response and participation.) On the other hand, many AM, FM, TV and domestic SW broadcasters couldn't care less if they are heard outside their intended coverage area. And many utility stations are not pleased at all to discover that someone other than the intended receiving station may have listened to one of their transmissions.

There are some solutions to these problems. One of the most common is for the listener to prepare a QSL card for the station, with all data filled in, which a station official can then sign and return to the SWL. Other QSL collectors enclose souvenirs, such as picture postcards, decals, or postage stamps, with their reports. While reports to international broadcasters can be in English (or any of the other languages they broadcast in), most foreign domestic SW broadcasters and utility stations in non-English-speaking nations will have no one who understands English; reports to such stations will have to be written in the station's language (often using a form available from various SWL clubs). To ensure that their

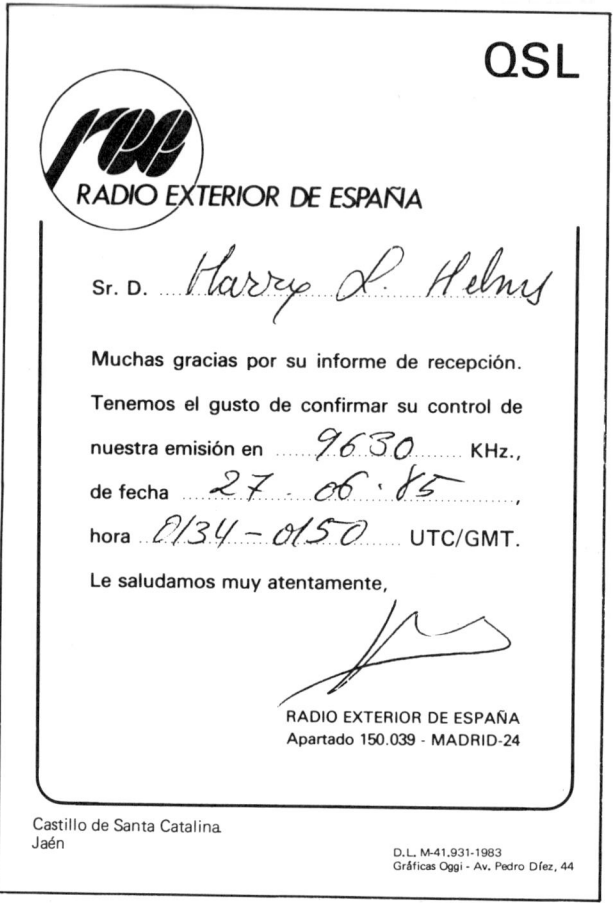

Figure 11.1 The goal of many SWLs and DXers—a QSL card such as this, was received from Spanish Foreign Radio, confirming reception.

reception report arrives at a station, and to give it the aura of importance, some listeners use registered mail if their first few reports go unanswered. Patience is often needed to finally extract a QSL from many stations; some hardcore QSL collectors have sent more than a dozen reports to a single station over a period of several years in hopes that someone will finally verify their reports.

The basic elements of any reception report are the following:

- the frequency on which you heard the station
- the month, day, and year you heard the station
- the time your reception began and ended
- the quality of the station's signal, including strength and QRM

- a brief description of your receiver and antenna
- if the station is an international broadcaster, some comments and reactions to their programs
- a *request* for the station to verify your report if it is correct
- enough details of what you heard to prove that you actually heard the station

The frequency on which you heard a station is simple to determine if you have a receiver equipped with direct-frequency readout; otherwise, you'll have to rely upon announcements by the station. Sometimes, you'll note that the announced frequency does not agree with the received frequency, as when Radio Moscow announces a rounded frequency such as "9.53 MHz" when they're transmitting on 9535 kHz. The best policy in such cases is to report both the announced and actual frequencies.

The date and time items are interrelated. If you are reporting to an international broadcaster or utility station, the time should *always* be in UTC even if the station itself gives times in the local time of its intended audience areas. If you are reporting to a domestic station, the best bet is to report in the local time where the station is located. For example, if you're in the Mountain time zone in North America and hear an AM or FM station in the Eastern time zone, you should report reception using Eastern time. Moreover, the date you should use in your reception report depends upon the time in which you report. If you hear a European station during the evening in North America, it is the next day in UTC—and the date you specify for the reception should be the next day, not your local date.

Indicate the beginning and ending times of your reception. How long a period to listen depends on several factors. If you're reporting to a major international broadcaster, a period of approximately a half-hour is a good target. (In fact, some broadcasters specify minimum periods that one must listen to earn a QSL.) In other cases, reception will of necessity be brief. This often occurs during sporadic-E or other short-lived propagational openings; most utility transmissions are short by their very nature (and there's no use in listening long to a repeating marker transmission). And there's a limit to what you can report if a station signs off soon after you first tune in.

How to describe the station's signals can be difficult. For years, many SWLs have used the "SINPO" code to indicate how well a station was received. "SINPO" is an acronym for *S*ignal strength, *I*nterference, *N*oise, *P*ropagation, and Overall quality. The scale used ranges from 1 to 5, with 5 denoting the best possible condition (loudest signal, no interference, no noise, excellent propagation, and superb overall quality) and 1 representing a condition so poor as to make the signal unusable or unlistenable. Although it is a convenient shorthand, the SINPO code suffers from the unavoidable fallacy found whenever an "objective" number is assigned to an inherently subjective evaluation; a signal that is "S4" to one listener would be "S3" to another, while a "P2" propagation rating would be judged "P3" by another SWL. This also shows up in some SWL reports containing clearly illogical SINPO ratings, such as "SINPO 32325." Despite this, some international broadcasters still promote the use of SINPO, as do a few SWL clubs.

Your author suggests that more extensive written descriptions of signal quality be used instead. A description such as "your signals were of good to very good strength, with only slight, regular fading, but with heavy interference at times from the BBC relay at Antigua on 6185 kHz" conveys far more useful information than "SINPO 43543." Moreover, SINPO does not allow for the dynamic nature of SW reception; the written description allows the SWL to indicate that the QRM from the BBC was present only at different times during the reception instead of constantly and that the received signal strength varied. A description of noise was omitted entirely, since in most cases it is generated by the SWL's local environment; if a SWL has noise from a power transformer or neon sign down the street, what can the station do about it?

A final argument against the SINPO code is that it is seldom understood by other than international broadcasters. Domestic broadcasters (including AM, FM, and TV stations) and utilities usually have no idea what is meant by "SINPO 43544," so a written description is your only alternative in reporting those stations.

If a station is interested in the fact that you heard them, they will also be interested in your receiver and antenna. There is no need to go into an exhaustive description of every item you have or an extensive technical review of your receiver. You can cover this point easily by giving the brand and model number of your receiver along with a brief descriptive phrase in case the station isn't familiar with it, such as "portable receiver" or "communications receiver." If you had to use a special feature or technique to receive the station, such as ECSSB, a narrow bandwidth filter, or preamplifier, mention this as well. Describe your antenna as a longwire, dipole, and so on, and if you give its dimensions (length, height above ground, and the like), do so in meters and centimeters rather than feet and inches.

It's difficult to comment on the "programs" of a utility station, and it's not a good idea at any rate. You can be more expressive when it comes to the programming of domestic SW broadcasters, even if you don't understand the language. For example, you could remark that you enjoyed listening to the station because of its music or because you are interested in learning the language of that nation. But major international broadcasters are becoming increasingly interested in comments on their programming, judging from the remarks of personnel at many such stations which have appeared in SWL publications. The implication of such remarks seems to be that SWLs have favorable comments and useful suggestions to make about programs, but don't in their haste to secure a QSL. Moreover, some in the SWL hobby have implied that it's crass to listen to a station simply for a QSL alone. Such stations are on the air to be listened to, so goes this argument, and any SWL who listens to a station just to secure a QSL is shirking his or her responsibility to the station. However, it never seems to occur to some station personnel or some factions of the SWLing hobby that most normal people might be bored to tears by the programming of some stations and indeed listen only for a QSL.

However, your author feels that remarks about programming are an essential part of reports to international broadcasters, as long as those remarks are *honest* and *candid*. If you couldn't care less about the growth in a particular country's steel output, say so. If you feel that Radio Moscow needs to be reminded that Americans can get news about the United States from sources other than Tass, remind them. If you feel a station is a waste of its nation's money and time, say so. If you would like more programs about a nation's history or ethnic groups, ask for more. If you enjoy music from a particular nation, tell them. And say some kind words about announcers you like—or unkind words about those you don't. Some SWLs are tempted to say nice things about a station's programs just to improve their chances to get a QSL or a "goodie" such as a station pennant. This, unfortunately, only perpetuates mediocre programming practices.

Some stations, particularly from the Soviet bloc nations, will take listener comments and remarks out of context. The Soviet bloc is not the only offender in this regard; as noted earlier in this book, South Africa's Radio RSA has been known to pull similar stunts. However, Soviet bloc stations go out of their way to stimulate remarks which can be used in ways other than those the SWL intended. This doesn't mean that you should avoid commenting on programs from such stations, but your observations should be clear and unambiguous.

It's also important to include a request that the station verify your report if it is correct. One reason is that some stations will send a QSL only upon request. A pet phrase your author usually includes is, "If this reception report is correct and of use to your station's staff, I would greatly appreciate receiving a card or letter confirming that I indeed heard your station at the time, date, and frequency noted in this letter." Some SWLs demand a QSL and wonder why they sometimes receive only a program schedule in response; station personnel, like everyone else, usually respond better to requests than ultimatums. If you're reporting to a station such as a utility, you may find that it's effective to explain why you want a QSL from that station, such as, "My hobby is trying to hear as many different shortwave (or AM, FM, and so on) stations as I can, and to collect cards and letters from these stations confirming that I did indeed hear them. Thus, a card or letter from your station would be a valued addition to my collection."

The major part of your reception report will likely be devoted to details and information to prove that you indeed heard the station. Admittedly, many stations do not fully check reports, and it's not unheard of for a listener to receive a "QSL" in response to a request for a program schedule. However, some stations do carefully check reports, and some stations retain reports in their files or release them to local SWL clubs (for prospective members, and the like). Some SWLs have unfortunately developed reputations as "reception fakers," persons whose

Dear Listener,

The North American Service of Radio Moscow invites listeners in the United States and Canada to send their comments on the new Soviet disarmament proposals outlined in the statement of the General Secretary of the Central Committee of the Soviet Communist Party Mikhail Gorbachov on January, the 15th. If you send your opinions in written or tape form you'll be able to hear them on our program "Opinion Forum".

What do you think of the Soviet proposal to rid the planet of the nuclear weapons by the end of the century under strict control that would include on-site verification of the way nuclear warheads and carriers are destroyed or dismantled.

What do you think of the Soviet proposal to leave outer space peaceful, not to develop or station strike weapons there and establish strict control, among other things to open appropriate laboratories for inspection.

What do you think of the Soviet Union's proposal to mutually stop all nuclear explosions, introducing reliable verification, including on-site inspection?

Send your written or recorded opinions on these and other Soviet proposals to the North American Service of Radio Moscow, Moscow, USSR. The Soviet Union stands for entering the third millenium without nuclear or space weapons, as well as all other means of mass destruction. This concerns the survival of humanity. This concerns every person, no matter where he or she might live.

Figure 11.2 Requests for comments and opinions, such as this from Radio Moscow, can be a trap for the naive SWL.

impressive QSL collections are based upon fraudulent reports. It's possible to "fake" receptions based upon material appearing in SWLing publications, but the damage to a SWL's reputation can be (and usually is) lasting if one of the fraudulent reports circulates within the SWL hobby.

The usual format is to list each item you are quoting in order by the time at which you heard it. The best possible material to prove your reception are items that you could not have possibly known about without actually having heard the station. For example, did the station sign off or on at a different time than scheduled? Did the station use a new frequency (perhaps as a result of having drifted off its assigned one?) or experience unusual technical difficulties? Was there a sudden, unexpected change in normally scheduled programming?

The next best items, particularly if you are reporting normally scheduled programs, are things such as the names of announcers, titles of musical selections played, items given in a newscast, topics of any commentaries and news analyses, names of listeners whose letters were read or who requested musical selections, persons interviewed on the air, and the like. Advertisements are excellent items to quote in reporting reception of commercial stations. The least convincing items to report are those which could be gathered from program schedules and SWL publications. Remember, the burden is upon you to prove that you heard the station. Using items that are "public knowledge" or vague does not help your cause.

The tendency of some stations to send out QSLs indiscriminately means that you should never rely on a station's QSL to prove that you heard a station *if* you are not sure yourself that you actually heard it. Some SWLs send out "tentative" reports which, in effect, ask the station "did I really hear you?" This technique is acceptable if you have strong reason to believe that you did indeed hear the station, but shouldn't be relied on if there are two or more possible stations you could have heard and you're not sure which one it was.

The entire matter of whether a QSL "proves" reception is often hotly debated in the SWL hobby, particularly among those SWLs and DXers who engage in the more competitive aspects. It is clear that a QSL by itself does not provide such proof; whether or not a SWL's report of hearing a rare DX station is believed will depend more upon that SWL's reputation as an accurate, reliable, and honest reporter of what he or she heard than it will upon whether or not a station will QSL the report.

Postage is a consideration in reporting reception of a foreign station. Airmail is more expensive than surface mail, but the alternative may be for the station to receive your report three or four months after it was mailed. Airmail ensures that the station will receive your report in time for it to be useful. Your airmail postage costs can be minimized by buying onion-skin envelopes and paper for your reports. If you don't need to add any enclosures to your report, *air letter sheets* or *aerogrammes* can cut your postage bill considerably. Available from any post office, these are sheets of paper with preprinted postage. You write your letter on one side of the aerogramme and then fold it into the shape of an envelope and seal it by the gummed flaps provided. Aerogrammes have room for a complete reception report, and the cost is less than a conventional airmail letter sent in a envelope. If you report major international broadcasters, aerogrammes will usually be the most cost-effective way to do so. However, if you enclose postcards, stamps, a prepared QSL, or similar items, you'll have to use a regular airmail envelope.

Major international broadcasters will send out QSLs without the listener's having to supply return postage. Some, however, require return postage if you want the QSL sent airmail. Domestic and private SW stations, as well as some international broadcasters operated by less affluent nations, do appreciate (and sometimes require) return postage for the QSL. The most popular method of providing return postage is through one or more *International Reply Coupons* (IRCs), which are currently available at most major post offices. An IRC can be exchanged at post offices in virtually any country of the world for postage stamps to pay for a *surface* mail letter to a foreign country; thus, you'll need to send two or three IRCs to pay for a QSL via airmail.

However, IRCs are not always the best way to prepay QSL postage. In many nations, IRCs may not be well understood by the local post offices, and it's not unusual for them to refuse to redeem IRCs. Moreover, the post office in some areas can be quite a distance from the station; it might take a day to go to the post office and back. IRCs are particularly unproductive when sent with prepared QSLs, since the object is to make it as easy as possible for the station to verify your report. The solution is such cases is to send mint stamps of the country concerned with your report. Such mint stamps can be obtained from stamp dealers or from the "DX Stamp Service," a commercial venture which sells the appropriate amount of foreign postage stamps to return a post card or one-half ounce letter by airmail to the United States or Canada. Their address is 83 Roder Parkway, Ontario, New York 14519; their latest price list is available for a stamped, self-addressed envelope (SASE). In many cases, sending mint stamps is a cheaper way to pay for an airmail reply than IRCs.

A final method to increase the effectiveness of your reception reports is to typewrite or carefully print your reports. Most handwriting is difficult enough to read already, and imagine the problems faced by station personnel whose first language may not be English.

RECORD KEEPING AND AWARDS

Many SWLs and DXers have a habit of keeping track of everything they hear in a written record known as a *log*. A log lets you keep track of what you've heard, and some DXers can instantly rattle off the number of different stations and countries they've heard and verified. These are known in the hobby as "totals."

The exact form a log can take varies with the preferences of the individual listener. Some SWLs use printed logging sheets and forms, which are available from SW equipment dealers, while others use ruled notebooks similar to those used by students. The amount of information included also depends upon the preferences of the SWL. Some are content with merely recording the call, frequency, and time of a station, while others record virtually every detail of each reception.

This sort of record keeping is encouraged by some clubs, who publish rankings of DXers by the number of stations and countries heard and verified. In addition, some clubs offer awards (usually in the form of a certificate) for collecting QSLs from specified numbers of countries (such as 100 or 150 countries) along with regional awards for verifying such things as the various USSR republics, Nigerian regional stations, or all countries in a continent. A written log is important in such cases for determining which countries are still needed for an award.

SWLs who don't care about awards or rankings can find a log useful. For example, it can be a good tool for determining reception patterns from different parts of the world at different seasons of the year; if K-index and solar flux values are also included in the log, it can serve as a guide to what can be expected at a certain time of year, given approximate values of the K-index and solar flux. Other SWLs make their log into a diary of sorts, complete with observations and remarks about the general condition of the SW bands, results of comparisons between different antennas and equipment, and so on.

Of course, many (if not most) SWLs don't keep a log, nor do they care about totals, rankings, or awards. Like so much of the SWL hobby, the choice of whether to keep a log or participate in the more competitive aspects is strictly a matter of individual choice.

SWL "CALL SIGNS" AND IDENTIFIERS

Back in the late 1950s, the editors of the now-defunct *Popular Electronics* magazine came up with the concept of assigning "identifiers," similar to amateur call signs, to SWLs who desired them. While new to the American SWL scene, this practice had been going on for some time in Europe (particularly Soviet bloc nations) and was usually intended for those SWLs who aspired to become hams; such SWLs were issued identifiers by the national amateur radio organizations so that they could feel like "one of the gang" before actually getting licensed. The editors of *Popular Electronics* decided to begin their identifiers with the prefix "WPE," followed by a digit (corresponding to the amateur radio call sign districts) and three more letters. Not only did this arrangement manage to avoid conflict with any actual call sign assignments, it also was a good advertisement for the magazine.

Thousands of these "WPE monitoring certificates" were issued by *Popular Electronics* and soon SWLs began to include their WPE "call letters" in correspondence to stations and the magazine. Not that these calls and certificates were handed out to just anyone; you had to have received five QSLs, with at least one from a station outside the United States—and don't forget the 25c in coins (no stamps, please)! Some SWLs even had their own "QSL" cards printed with their calls, and soon were busy swapping cards with other SWLs and including their calls in their reception reports. By the mid-1960s, it seemed as if every self-respecting SWL had a WPE "call" and "QSL" cards to boot.

By the late 1960s, *Popular Electronics* decided to decrease the amount of SWL-related material in the magazine, including discontinuing the monthly SWL column and the entire WPE program. The magazine's SWL column editor took over what remained of the WPE program, and continued issuing identifiers beginning with the "WDX" prefix. Since the WPE/WDX program was mainly intended for SWLs listening to 30 MHz and below, another registration program for scanner listeners was started by a company specializing in frequency guides and directories for those listeners. Prefixes in this series begin with "K" followed by the post office two-letter abbreviation for the listener's home state. But deprived of the driving engine of *Popular Electronics* the entire business of SWL calls and card swapping soon faded. Today, it is still practiced but by only a fraction of the numbers who participated when the craze was at its peak. If you do run across what appears to be a call sign beginning with "WDX2" or "KFL4," however, you'll know it's a SWL or scanner listener and not an actual transmitting station.

Many SWLs who were around during the height of the WPE mania have fond memories of that period, and one SWL club devoted to WPE calls and other 1960s SWLing nostalgia was formed. Indeed, it sometimes seems as if the hobby was a little more fun in that era, when many SWLs didn't take themselves or the hobby quite so seriously. (And, yes, your author was proud to be known as WPE4HKE back in that period!)

SWL CLUBS

SWL clubs are nonprofit associations of SWLs who band together to exchange information and tips regarding their particular listening interests. All labor is voluntary and unpaid, and the work load of publishing a monthly (or more frequent) club bulletin and administering club affairs is often a heavy burden. For most club officers and workers, the only reward they derive is the satisfaction of knowing that they do a good and vital job or, if they actually publish the bulletin, in getting a copy of the bulletin before anyone else. The various columns and features in club bulletins are put together by unpaid editors, who donate their time and efforts for the benefit of the rest of the club. All clubs charge dues for membership, but these dues are almost exclusively consumed by the expenses of printing and mailing the bulletin as well as club management and administration expenses. (There is an almost perfect correlation between increases in postage rates and increases in club membership dues.) If one wants to get rich, starting and running a SWL club is not the way to do it!

Most club bulletins today are printed by using the "offset" technique; this results in a bulletin printed on an 8.5 x 11-in. sheet, with two bulletin pages printed on the front and back sides for a total of four bulletin pages per sheet. The sheets are then folded and stapled to produce a "booklet" style bulletin which can be easily mailed and read. All material is prepared by using typewriters or computer printouts; the offset printing techniques allows printing plates to be made from such input and also allows illustrations of QSL cards to be reproduced; photographs can also be reproduced but with reduced quality. A few bulletins, primarily from smaller clubs, are still prepared by using mimeograph or similar reproductive techniques.

The subject of club management is often a source of controversy. Some clubs hold elections for officers to develop club policies and manage the club; the publication of the bulletin is in the hands of a separate publications group. Most clubs, however, are managed by the same group that publishes the bulletin; the principle here is that those who do the work, which is unpaid, should have the final say in running the club. This does not mean that such clubs are always dictatorships or totalitarian. Almost all such clubs are very responsive to member needs and concerns, with members often asked to write in their opinions on certain issues or to respond to polls. The final test is how well a club serves the membership, and it is often the case that "nondemocratic" clubs serve their memberships better than "democratic" ones.

It is also true that in many cases there is no real alternative to allowing the publishers of the bulletin to control the club. The amount of effort required to publish a bulletin is enormous, and without a bulletin there is no club. In fact, the most common reason why SWL and DX clubs go out of existence is not a lack of members but an inability to find someone willing to take over responsibility for publishing the bulletin when the existing publisher resigns. Finally, most SWLs and DXers are "apolitical," in that club management and control are largely irrelevant to them as long as the club bulletin and other programs enhance their listening.

SWLing and DXing, and the clubs serving participants, have, like all human activities, certain norms and traditions which can be baffling to an outsider or newcomer. Indeed, the "mix" of factors potentially involved, such as international politics, technical expertise, and competition for new QSLs, makes for a culture that could probably keep an army of sociologists and psychologists gainfully employed for years studying it.

Throughout this book, reference has been made to "SWLing" and "DXing" as two separate activities. Years ago, the distinction between these two was not as great because any SWLing was, in effect, DXing as well; even the BBC or Radio Moscow took some specialized knowledge and skill to receive reliably. That changed with the advent of "superpower" transmitters and more advanced receivers, which were easier to use. SWL clubs have likewise reflected this change. In 1970, the orientation of almost all SWL clubs was toward DXing. At the time this book was written, most clubs have a heavy emphasis on SWLing as opposed to DXing. In fact, many DXers interested in the more competitive aspects, such as hearing and verifying new countries and rare stations, have formed their own organizations in which membership is available only by invitation; the object of this policy is to "weed out" those whose interests lie outside hard-core DXing. The name of one such group (considered to be the most elite of all) perhaps unintentionally drives this point home—it's called "Numero Uno," which is Spanish for "number one."

Needless to say, this has lent a certain status to being a member of one of these DXing "secret societies." It has also resulted in conflict, with SWLs charging such groups with "hoarding" information for the private benefit of their members, while many DXers feel that many SWL-oriented clubs are unresponsive to their needs. In fact, hard-core DXers have a point; in the late 1970s, the traditional DXing hobby was often vilified in some SWL club bulletins, as members were exhorted to listen to stations for their programming and eschew QSLs, totals, and other competitive aspects of the hobby. Those interested in DXing often had little choice but to form their own organizations to serve their needs. (One of the more curious aspects of this entire business was that one of the more vocal proponents for the new SWLing emphasis started his campaign in earnest after his membership in Numero Uno was cancelled for failing to supply enough DX-related material to its bulletin.)

Fortunately, the average SWL or DXer doesn't have to worry about such shenanigans. Many SWL clubs still have enough DXing material to satisfy you if you get interested in chasing rare stations and QSLs. However, it must be admitted that your goal, should the DXing bug bite hard, ought to be to get invited to join one of the "invitation-only" DXing groups. There are more practical reasons for this than just "prestige." One important one is that such groups publish their newsletters on a weekly or biweekly basis; the information is much "fresher" than in monthly SWL club bulletins. This can be important if the information involves openings present only briefly each year or off-frequency operation and new stations. Another good reason is that most serious DXers now concentrate their activity in such groups. If you want advice and tips from the best, you need to be part of one of those groups as well. The way to get an invitation is to report your best DX each month to SWL club bulletins. If your catches are considered good, and your reputation as an accurate and honest reporter of DX information builds, one day an invitation to join a "private" DX group will arrive in your mailbox.

Which SWL club (or clubs) should you join? There is no easy way to answer to that, since each club has its own particular (and sometimes peculiar) personality and listening emphases. For example, some clubs are "all band," meaning they cover everything from AM DXing to SWLing to FM/TV DX. Others specialize in segments of the hobby such as AM DXing, clandestine and pirate stations, or LW reception. Moreover, the strengths and weaknesses of clubs change over time, depending on the editors and members involved. It is not unknown for a strong, vital club to have its bulletin rapidly deteriorate and its membership drop following the resignation of key editors or club officers. By the same token, a moribund club can be revitalized by new editorial talent.

Some clubs also have a tendency to get involved in areas that have little, if any, relevance to SWLing and DXing or disintegrate into soapboxes for the political and social opinions of various editors and club officials. For others, being a club official or editor provides their first taste of "power" and "status," resulting in their fragile egos running amok. Childish disputes can erupt and reverberate for years. (Your author remembers the case of one DXer, expelled as a teenager from a club, who—six years later and then a college senior—sent copies of letters of acceptance received by him from various law schools to his former "enemies," apparently in an attempt to show them how "wrong" they were about him!) Other SWLs seem to enjoy stirring up controversies within certain clubs; often, the motivation seems to be their need for attention at any cost and a lack of anything better to do with their time and energy. Sadly, such conflicts rarely concern substantive matters, but instead are pure personality clashes. Fortunately, there is no need for any SWL or DXer to concern himself or herself with such "trivial pursuits"—or to support a club or commercial publication that allows such irrelevant nonsense to take up space which could otherwise be filled with SWLing and DXing news.

The only way for you to judge whether or not a club might be for you is to examine a sample copy of its bulletin. Fortunately, almost every club will be willing to send you a sample bulletin and membership information. Since clubs are nonprofit ventures, it is best to enclose one dollar to cover the costs of the sample bulletin and mailing. A list of the current addresses of major SWL clubs is included in the appendix of this book. The major SWL and DX clubs in North America have also joined together in an organization known as the Association of North American Radio Clubs (ANARC). ANARC serves as a clearinghouse for news about clubs and their activities and also sponsors an annual convention which brings together SWLs, DXers, broadcasting personnel, and equipment manufacturers from all over the world. ANARC has a list of clubs which are currently ANARC members. It can be obtained for a stamped self-addressed envelope sent to ANARC at 1500 Bunbury Drive, North Whittier, California 90601.

Once you find clubs which agree with your needs and interests, you'll find them valuable adjuncts to your listening. One aspect that many like is the opportunity for contact with other SWLs. It is always interesting to see the kind of reception others in your general area or those using similar equipment are able to achieve. Some clubs allow members space for their opinions and observations, and this can be an important part of getting to know other members better. (It can also, unfortunately, lead to some of the problems cited earlier.) Most clubs also permit members to buy and sell receivers and other equipment through free ads in the bulletin.

Another way some SWLs promote greater contact is by organizing regional groups, with membership open only to residents of certain geographic areas. Some of these groups issue their own club bulletins and allow SWLs outside the intended area to subscribe to the bulletin. Others are primarily social in nature, and have regular meetings at which members can exchange the latest news and listening tips or compare QSLs.

SWL-to-SWL contact has also entered the computer age through the establishment of various "bulletin boards" which can be accessed by those who have a microcomputer and modem for telephone line communication. These bulletin boards allow tips and information to be exchanged in a matter of minutes with other listeners, and will likely become an increasingly important information source for active SWLs.

The degree to which you decide to participate in SWL club activities, other aspects of the hobby, or even whether you want to consider SWLing a hobby is up to you. As mentioned in Chapter 1, SWLing is what you want it to be and what you make it. It is an intensely personal venture; you can explore those interests which appeal to you—whether they be regular listening to a few stations, seeking out rare DX stations, or even SWL club politics—without having to seek anyone's permission or approval. You can add or drop interests and activities as you please, and be as private or public as you wish about it. The choices are yours, and the only guideline is that your decision maximize your listening enjoyment.

The world is out there, saying many things in many different ways. The magic box that is a SW radio will let you hear it all.

APPENDIX

CALL SIGN ALLOCATIONS OF THE WORLD

AAA–ALZ	United States
AMA–AOZ	Spain
APA–ASZ	Pakistan
ATA–AWZ	India
AXA–AXZ	Australia
A2A–A2Z	Botswana
A3A–A3Z	Tonga
A4A–A4Z	Oman
A5A–A5Z	Bhutan
A6A–A6Z	United Arab Emirates
A7A–A7Z	Qatar
A9A–A9Z	Bahrain
BAA–BZZ	Peoples Republic of China (BVA–BVZ used by Taiwan)
CAA–CEZ	Chile
CFA–CKZ	Canada
CLA–CMZ	Cuba
CPA–CPZ	Bolivia
CQA–CUZ	Portugal and its territories
CVA–CXZ	Uruguay
CYA–CXZ	Canada
C2A–C2Z	Nauru
C3A–C3Z	Andorra
C4A–C4Z	Cyprus
C5A–C5Z	The Gambia
C6A–C6Z	Bahamas

C8A–C9Z	Mozambique
DAA–DTZ	Germany (DMA–DMZ and DTA–DTZ used by East Germany)
DUA–DZZ	Philippines
D2A–D3Z	Angola
D4A–D4Z	Cape Verde Islands
D6A–D6Z	Comoros
EAA–EHZ	Spain and its territories
EIA–EJZ	Ireland
EKA–EKZ	USSR
ELA–ELZ	Liberia
EMA–EOZ	USSR
EPA–EQZ	Iran
ERA–ERZ	USSR
ESA–ESZ	Estonian SSR
ETA–ETZ	Ethiopia
EUA–EWZ	Byelorussian SSR
EXA–EZZ	USSR
GAA–GZZ	United Kingdom
HAA–HAZ	Hungary
HBA–HBZ	Switzerland
HCA–HDZ	Ecuador
HEA–HEZ	Switzerland
HFA–HFZ	Poland
HGA–HGZ	Hungary
HHA–HHZ	Haiti
HIA–HIZ	Dominican Republic
HJA–HKZ	Colombia
HLA–HMZ	South Korea
HNA–HNZ	Iraq
HOA–HPZ	Panama
HQA–HRZ	Honduras
HSA–HSZ	Thailand
HTA–HTZ	Nicaragua
HUA–HUZ	El Salvador
HVA–HVZ	Vatican City
HWA–HYZ	France and its territories
HZA–HZZ	Saudi Arabia
IAA–IZZ	Italy and administered territories
JAA–JSZ	Japan
JTA–JVZ	Mongolia
JWA–JXZ	Norway
JYA–JYZ	Jordan
J2A–J2Z	Djibouti
J3A–J3Z	Grenada
J5A–J5Z	Guinea–Bissau
J6A–J6Z	St. Lucia
J7A–J7Z	Dominica
J8A–J8Z	St. Vincent
KAA–KZZ	United States
LAA–LNZ	Norway
LOA–LWZ	Argentina
LXA–LXZ	Luxembourg
LYA–LZZ	Lithuanian SSR

Call Sign Allocations of the World

L2A–L9Z	Argentina
MAA–MZZ	United Kingdom
NAA–NZZ	United States
OAA–OCZ	Peru
ODA–ODZ	Lebanon
OEA–OEZ	Austria
OFA–OJZ	Finland
OKA–OMZ	Czechoslovakia
ONA–OTZ	Belgium
OUA–OZZ	Denmark and its territories (including Greenland)
PAA–PIZ	Netherlands
PJA–PJZ	Netherlands Antilles
PKA–POZ	Indonesia
PPA–PYZ	Brazil
PZA–PZZ	Surinam
P2A–P2Z	Papua New Guinea
P5A–P5Z	North Korea
QAA–QZZ	Reserved for international "Q" signals
RAA–RZZ	USSR
SAA–SMZ	Sweden
SNA–SRZ	Poland
SSA–SSM	Egypt
SSN–STZ	Sudan
SUA–SUZ	Egypt
SVA–SZZ	Greece
S2A–S3Z	Bangladesh
S6A–S6Z	Singapore
S7A–S7Z	Seychelles
S8A–S8Z	Transkei district of South Africa
S9A–S9Z	Sao Tome e Principe
TAA–TCZ	Turkey
TDA–TDZ	Guatemala
TEA–TEZ	Costa Rica
TFA–TFZ	Iceland
TGA–TGZ	Guatemala
THA–THZ	French territories
TIA–TIZ	Costa Rica
TJA–TJZ	Cameroon
TKA–TKZ	French territories
TLA–TLZ	Central African Republic
TMA–TMZ	French territories
TNA–TNZ	Congo
TOA–TQZ	France
TRA–TRZ	Gabon
TSA–TSZ	Tunisia
TYA–TYZ	Benin
TZA–TZZ	Mali
T2A–T2Z	Tuvalu
T3A–T3Z	Kiribati
UAA–UZZ	USSR
VAA–VGZ	Canada
VHA–VNZ	Australia
VOA–VOZ	Canada

VPA–VSZ	British colonies and administered areas
VTA–VWZ	India
VZA–VZZ	Australia
V2A–V2Z	Antigua and Barbuda
V3A–V3Z	Belize
V4A–V4Z	St. Christopher
V8A–V8Z	Brunei
WAA–WZZ	United States
XAA–XIZ	Mexico
XJA–XOZ	Canada
XPA–XPZ	Denmark and its territories (including Greenland)
XQA–XRZ	Chile
XSA–XSZ	China
XTA–XTZ	Burkina Faso
XUA–XUZ	Cambodia (Kampuchea)
XVA–XVZ	Vietnam
XWA–XWZ	Laos
XXA–XXZ	Portuguese territories
XYA–XZZ	Burma
YAA–YAZ	Afghanistan
YBA–YHZ	Indonesia
YIA–YIZ	Iraq
YJA–YJZ	New Hebrides
YKA–YKZ	Syria
YLA–YLZ	Latvian SSR
YMA–YMZ	Turkey
YNA–YNZ	Nicaragua
YOA–YRZ	Romania
YSA–YSZ	El Salvador
YTA–YUZ	Yugoslavia
YVA–YVZ	Venezuela
YZA–YZZ	Yugoslavia
ZAA–ZAZ	Albania
ZBA–ZJZ	British territories
ZKA–ZMZ	New Zealand
ZNA–ZOZ	British territories
ZPA–ZPZ	Paraguay
ZQA–ZQZ	British territories
ZRA–ZUZ	South Africa
ZVA–ZVZ	Brazil
Z2A–Z2Z	Zimbabwe
1AA–1AZ	Unofficially used in disputed territories such as the Spratly Islands
2AA–2ZZ	United Kingdom
3AA–3ZZ	Monaco
3BA–3BZ	Mauritius
3CA–3CZ	Equatorial Guinea
3DA–3DM	Swaziland
3DN–3DZ	Fiji
3EA–3FZ	Panama
3GA–3GZ	Chile
3HA–3UZ	China
3VA–3VZ	Tunisia
3WA–3WZ	Vietnam

Call Sign Allocations of the World

3XA–3XZ	Guinea
3YA–3YZ	Norway
3ZA–3ZZ	Poland
4AA–4CZ	Mexico
4DA–4IZ	Philippines
4JA–4LZ	USSR
4MA–4MZ	Venezuela
4NA–4OZ	Yugoslavia
4PA–4SZ	Sri Lanka
4TA–4TZ	Peru
4UA–4UZ	United Nations
4VA–4VZ	Haiti
4WA–4WZ	Yemen
4XA–4XZ	Israel
4YA–4YZ	International aviation
4ZA–4ZZ	Israel
5AA–5AZ	Libya
5BA–5BZ	Cyprus
5CA–5GZ	Morocco
5HA–5IZ	Tanzania
5JA–5KZ	Colombia
5LA–5MZ	Liberia
5NA–5OZ	Nigeria
5PA–5QZ	Denmark and its territories
5RA–5SZ	Malagasy Republic
5TA–5TZ	Mauritania
5UA–5UZ	Niger
5VA–5VZ	Togo
5WA–5WZ	Western Samoa
5XA–5XZ	Uganda
5YA–5ZZ	Kenya
6AA–6BZ	Egypt
6CA–6CZ	Syria
6DA–6JZ	Mexico
6KA–6NZ	South Korea
6OA–6OZ	Somalia
6PA–6SZ	Pakistan
6TA–6UZ	Sudan
6VA–6WZ	Senegal
6XA–6XZ	Malagasy Republic
6YA–6YZ	Jamaica
6ZA–6ZZ	Liberia
7AA–7IZ	Indonesia
7JA–7NZ	Japan
7OA–7OZ	Yemen Democratic Republic
7PA–7PZ	Lesotho
7QA–7QZ	Malawi
7RA–7RZ	Algeria
7SA–7SZ	Sweden
7TA–7YZ	Algeria
7ZA–7ZZ	Saudi Arabia
8AA–8IZ	Indonesia
8JA–8NZ	Japan

8OA–8OZ	Botswana
8PA–8PZ	Barbados
8QA–8QZ	Maldives
8RA–8RZ	Guyana
8SA–8SZ	Sweden
8TA–8YZ	India
8ZA–8ZZ	Saudi Arabia
9AA–9AZ	San Marino
9BA–9DZ	Iran
9EA–9FZ	Ethiopia
9GA–9GZ	Ghana
9HA–9HZ	Malta
9IA–9JZ	Zambia
9KA–9KZ	Kuwait
9LA–9LZ	Sierra Leone
9MA–9MZ	Malaysia
9NA–9NZ	Nepal
9OA–9TZ	Zaire
9UA–9UZ	Burundi
9VA–9VZ	Singapore
9WA–9WZ	Malaysia
9XA–9XZ	Rwanda
9YA–9ZZ	Trinidad and Tobago

INTERNATIONAL PHONETIC ALPHABET

Letter	Phonetic equivalent
A	Alpha
B	Bravo
C	Charlie
D	Delta
E	Echo
F	Foxtrot
G	Golf
H	Hotel
I	India
J	Juliet
K	Kilo
L	Lima
M	Mike
N	November
O	Oscar
P	Papa
Q	Quebec
R	Romeo
S	Sierra
T	Tango
U	Uniform
V	Victor

International Morse Code by Sound Chart

W	Whiskey
X	X–ray
Y	Yankee
Z	Zulu

INTERNATIONAL MORSE CODE BY SOUND CHART

Letter	Sound equivalent
A	Didah
B	Dahdididit
C	Dahdidahdit
D	Dahdidit
E	Dit
F	Dididahdit
G	Dahdidit
H	Didididit
I	Didit
J	Didahdahdah
K	Dahdidah
L	Didahdidit
M	Dahdah
N	Dahdit
O	Dahdahdah
P	Didahdahdit
Q	Dahdahdidah
R	Didahdit
S	Dididit
T	Dah
U	Dididah
V	Didididah
W	Didahdah
X	Dahdididah
Y	Dahdidahdah
Z	Dahdahdidit
1	Didahdahdahdah
2	Dididahdahdah
3	Didididahdah
4	Didididah
5	Dididididit
6	Dahdidididit
7	Dahdahdididit
8	Dahdahdahdidit
9	Dahdahdahdahdit
0	Dahdahdahdahdah
.	Didahdidahdidah
?	Dididahdahdidit
,	Dahdahdididahdah

Q-SIGNALS USED IN MORSE CODE COMMUNICATION

All signals can be made into questions by following each with a question mark.

Signal	Meaning
QRL	I am busy
QRM	Your transmission is being interfered with
QRN	I am troubled by static
QRO	Increase transmitter power
QRP	Decrease power
QRS	Send more slowly
QRT	Stop transmitting
QRU	I have nothing for you
QRV	I am ready
QRX	Call again
QRZ	You are being called by _____
QSL	I am acknowledging
QSO	I can communicate with _____
QSX	I am listening on _____
QSY	Change your frequency to _____
QTH	My location is _____

ABBREVIATIONS USED IN MORSE CODE TRANSMISSIONS

ABT	About
AGN	Again
ANT	Antenna
BK	Break
B4	Before
C	Yes
CK	Check
CL	Call
CQ	General call to any station listening
CUD	Could
CUL	See you later
DX	Distance; distant stations
ES	And
FB	Fine business (i.e., excellent)
GE	Good evening
GM	Good morning
GN	Good night
GND	Ground
HI	Laughter
HR	Here
HV	Have
LID	A poor or careless operator
NR	Number

OM	Old man (a general term used to refer to any male radio operator)
OP	Operator
OT	Old timer (veteran radio operator)
PSE	Please
PWR	Power
R	Received
RCVR	Receiver
RX	Receiver
SASE	Self-addressed stamped envelope
SIG	Signal
SKED	Schedule
SRI	Sorry
TNX	Thanks
TX	Transmitter
UR	Your (or you are)
VY	Very
WUD	Would
WX	Weather
XCVR	Transceiver
XMTR	Transmitter
XYL	Wife
YL	Young lady
73	Best regards
88	Love and kisses

CLUBS FOR SHORTWAVE LISTENERS

American Shortwave Listeners Club 16182 Ballad Lane, Huntington Beach, CA 92649. All-band coverage; monthly bulletin.

Association of Clandestine Enthusiasts P. O. Box 46199, Baton Rouge, LA 70895. Exclusively clandestine and pirate radio along with numbers stations; monthly bulletin.

Association of DX Reporters 7008 Plymouth Rd., Baltimore, MD 21208. All-band coverage; monthly bulletin.

Canadian International DX Club 6815 12th Ave., Edmonton, Alberta, T6K 3J6, Canada. All-band coverage; monthly bulletin.

International Radio Club of America P. O. Box 26254, San Francisco, CA 94126. Exclusively AM DXing; bulletin published 34 times per year, monthly in the fall and winter BCB DX "season."

Longwave Club of America, 45 Wildflower Rd., Levittown, NY 19057. Exclusively longwave (below 540 kHz) coverage; monthly bulletin.

National Radio Club P. O. Box 118, Poquonock, CT 06064. Exclusively AM DXing; bulletin published 30 times per year, weekly in the fall and winter BCB DX "season."

North American Shortwave Association 45 Wildflower Rd., Levittown, PA 19057. Exclusively SW broadcast; monthly bulletin.

Society to Preserve the Engrossing Enjoyment of DXing, 7738 East Hampton Court, Tucson, AZ 85715. Exclusively SW broadcast and utilities; monthly bulletin.

Worldwide TV-FM DX Association P. O. Box 514, Buffalo, NY 14205. Exclusively FM and TV DXing; monthly bulletin.

COMMERCIAL MONTHLY SWL PUBLICATIONS

Clandestine Confidential Newsletter RR #4, Box 110, Lake Geneva, WI 53147. Devoted exclusively to the latest information on clandestine broadcasters.

Monitoring Times, 140 Dog Branch Rd., Brasstown, NC 28902. Covers wide range of communications topics, with special emphasis on utilities.

Popular Communications, 76 North Broadway, Hicksville, NY 11801. Covers SWLing as well as telephones, scanners, radar detectors, etc.

Review of International Broadcasting, Box 490756, Fort Lauderdale, FL 33349. Heavy coverage of SW programming.

SHORTWAVE EQUIPMENT SUPPLIERS

CRB Research, P.O. Box 56, Commack, NY 11725. Frequency guides, directories, and related SWL publications.

Ege, Inc., 13646 Jefferson Davis Highway, Woodbridge, VA 22191. Receivers, accessories, and publications.

Electronic Equipment Bank, 516 Mill St. NE, Vienna, VA 22180. Receivers, accessories, and publications.

Gilfer Shortwave, 52 Park Ave., Park Ridge, NJ 07656. Receivers, accessories, and publications.

Radio West, 3417 Purer Rd., Escondido, CA 92025. Receivers, loop antennas for LW and the BCB, accessories, and publications.

Spectronics, 1009 Garfield St., Oak Park, IL 60304. Receivers, accessories, and publications.

Universal Shortwave, 1280 Aida Dr., Reynoldsburg, OH 43068. Receivers, accessories, and publications; sponsors free "on-line" bulletin board for owners of microcomputers and modems.

REFERENCE BOOKS FOR THE SHORTWAVE LISTENER

DXer's Directory, Universal Shortwave, 1280 Aida Dr., Reynoldsburg, OH 43068. Annual compilation of SWLs and DXers along with their addresses, listening interests, and club affiliations; listing is free and voluntary.

Reference Books for the Shortwave Listener

Radio Database International, International Broadcasting Services, Ltd., C.P.O. 300, Penn's Park, PA 18943. Computer-generated SW broadcasting schedules in graphic form with special coverage of tropical bands broadcasting; excellent receiver and equipment reviews.

Secrets of Successful QSLing, Tiare Publications, P. O. Box 493, Lake Geneva, WI 53147. Complete guide to reception reports and QSLing all types of radio stations written by well-known SWL Gerry Dexter.

Shortwave Clandestine Confidential, Universal Shortwave, 1280 Aida Dr., Reynoldsburg, OH 43068. Authoritative survey of clandestine broadcasting by Gerry Dexter, an acknowledged expert in the area.

The Beacon Guide, 6350 N. Hoyne Ave., Chicago, IL 60659. Guide to beacons operating on longwave, compiled by Ken Stryker.

Top Secret Registry of U.S. Government Frequencies, CRB Research, P. O. Box 56, Commack, NY 11725. Covers frequencies used by various government agencies for communications, although none (despite the title) are actually classified information. Essential for serious utility listeners.

World Radio TV Handbook, Billboard Publications, Inc., 1515 Broadway, New York, NY 10036. Annual directory of SW stations around the world; an indispensable reference for the active SWL.

INDEX

Accessories:
  audio filter, 79–80
  headphones, 80–81
  lightning arrestor, 68
  preamplifier, 77–78
  radioteletype (RTTY) receiving units, 81
Amateur ("ham") radio, 180–88
Antennas:
  active, 76–77
  dipoles, 68–73
  indoor, 74–75
  loop, 78–79, 175
  resonant, 62
  trap dipoles, 73
  tuning devices, 66
  vertical, 73–74
  1.6-30 MHz, 63-66

Broadcast band, standard AM (540-1699 kHz):
  clear channels, 170–73
  foreign stations, 174–77
  local channels, 170
  proposed expansion of, 176–77
  regional channels, 170
  sunrise/sunset reception patterns, 173
Broadcasters, domestic shortwave:
  Albania, 131
  Algeria, 138
  Angola, 138
  Argentina, 151
  Australia, 143–44
  Belize, 146
  Benin, 137
  Bolivia, 150
  Botswana, 136
  Brazil, 149–50
  Bulgaria, 132
  Burkina Faso, 137
  Burma, 141
  Canada, 145
  Central African Republic, 136–37
  Chile, 150–51
  China, 141
  Colombia, 149–50

241

Broadcasters, domestic shortwave: (*cont.*)
   Costa Rica, 146, 148
   Dominican Republic, 149
   East Germany, 131
   Ecuador, 149
   Falkland Islands, 146
   French Guiana, 147
   Gabon, 137
   Guatemala, 148
   Guyana, 146
   Haiti, 146
   Honduras, 146, 148
   India, 139–40
   Indonesia, 142–43
   Italy, 131–32
   Ivory Coast, 137
   Japan, 140
   Kuwait, 139
   Lesotho, 136
   Malaysia, 142
   Mexico, 148
   Mozambique, 138–39
   New Caledonia, 144–45
   Nicaragua, 148–49
   Nigeria, 134–36
   Pakistan, 140
   Paraguay, 150
   Peru, 150
   Saudi Arabia, 139
   Senegal, 137
   Singapore, 142
   Solomon Islands, 145
   South Africa, 133–34
   Southwest Africa (Namibia), 134
   Sri Lanka, 140
   Swaziland, 136
   Tahiti, 144
   Taiwan, 141–42
   Togo, 136
   Tunisia, 137–38
   Turkey, 139
   USSR, 132–33
   Venezuela, 149–50
   Vietnam, 142
   West Germany, 130–31
Broadcasters, international shortwave:
   British Broadcasting Co. (BBC), 100–103
   Deutsche Welle, 123–24
   HCJB, 115
   KCBI, 122
   KYOI, 121
   NDXE, 121
   Radio Beijing, 112–13
   Radio Canada International, 119–20
   Radio France Internationale, 110–12
   Radio Kiev, 107
   Radio Moscow, 103–108
   Radio Nederland, 108–110
   Radio RSA, 122–23
   Radio Station "Peace and Progress," 107
   Radio Tashkent, 107
   Radio Tirana, 114
   Radio Vilnius, 107
   Radio Yerevan, 107
   Trans World Radio, 116–17
   WINB, 120
   WRNO, 121
   Vatican Radio, 117–19
   Voice of America, 124–26

Clandestine broadcasting, 189–97

Emission modes:
   amplitude modulation (AM), 23–24
   continuous wave (CW), 22–23
   facsimile (FAX), 26
   frequency modulation (FM), 25–26
   radioteletype (RTTY), 26
   single sideband (SSB), 24–25

Index

slow-scan television (SSTV), 27

Filters:
audio, 52, 79–80
crystal, 49–50
high pass, 47
mechanical, 49–50
notch, 50–51
Frequency allocations, 27–39

"Numbers" stations, 202–207

"Pirate" broadcasting, 197–202
Propagation, radio wave:
auroral, 87, 91
beacon stations, 95–96
forecasting, 96–98
great circle path, 89
ground wave, 82
ionospheric structure, 84–85
K-index, 96–97
lowest usable frequency (LUF), 88
long haul path, 89
maximum usable frequency (MUF), 88
multihop, 90–91
multipath, 92
paths, 88–90
skip zones, 92–93
solar effects upon, 85–88
solar flux, 96–97
space wave, 83
sporadic-E, 84
sudden ionospheric disturbances (SIDs), 86–87

Receivers, shortwave:
automatic volume control (AVC), 53
beat-frequency oscillator (BFO), 55
choosing, 58–59
dynamic range, 47
frequency readout, 43–45
memories, 54
noise limiting circuitry, 52
S-meters, 53–54
selectivity: 48–52
sensitivity: 45–48
superheterodyne: 41–43
using, 57–58

Reception techniques:
exalted carrier single sideband (ECSSB), 55–57
gray line, 95

Shortwave listening (SWLing) hobby:
awards, 222–23
identifiers/"call signs," 223–24
"official" interest in, 210–13
reception reporting/QSLs, 215–22
record keeping, 222–23
SWLing clubs, 224–28, 237–38

Unidentified radio signals, 207–10
Utility stations:
aeronautical, 162–67
fixed, 158–61
longwave, 153–57
maritime, 157–58
standard time and frequency, 161–62

E1-